Drought Management and Socio-economic Development

Prof. Dr. T. S. Chouhan

Former Professor
Department of Geography
University of Rajasthan
Jaipur (India)

SCIENTIFIC® PUBLISHERS
New Delhi • Jodhpur

SCIENTIFIC®
PUBLISHERS

Published by:

Jodhpur
5 A, New Pali Road, P.O. Box 91
Jodhpur - 342 001 Rajasthan (India)
E-mail: info@scientificpub.com
Website: www.scientificpubonline.com

New Delhi
4806/24, Ansari Road, Daryaganj,
New Delhi - 110 002 (India)

ISBN: 978-93-89412-07-9 (H/B)
eISBN: 978-93-89412-09-3 (E/B)

Printed in India

Preface

The present book is most gigantic in view of widespread area of Rajasthan, largest state of the country with most adverse condition due to recurrence of drought. The drought has become recurrent feature as from 1950 till dates there were only 6 years where there was no drought in any part of the state.

Drought is one of the most severe disasters, which takes larger time for replenishment from the adversities. It is surprising at state rainfall variance is quite large varying from 18.55 cm in Jaisalmer to 95.03 cm in Banswara and all the districts have equal frequency of drought. The drought is treated as natural phenomena for failure of monsoon resulted in to crop damage of Kharif and Rabi seasons but severity of kharif crop remains severe in view of widespread area. As per criteria of the Government of India, Crop damage below 50 percent is treated as moderate condition, crop damage between 50 to 74 percent is treated as severe drought and crop damage from 75 to 100 percent is termed as most serious drought.

Failure of monsoon and occurrence of drought conditions create very serious problem to people as first and foremost problem before them is to survive family and livestock till next monsoon and expect State Government to start relief measures, which usually start from February till June. The drought of 2002 was most disastrous for being fifth consecutive drought years where relief measures were initiated from mid-August. Delay in relief measures causes large scale migration of people to cities and other parts of country in search of employment, survival of livestock becomes critical due to acute fodder and drinking water shortage. The economy of the state dwindles and loss vegetation on earth surface gets further loss.

Severity of drought is converted in to famine when it is combination of agricultural drought, hydrological and meteorological drought. There becomes acute shortage of food grain, fodder and drinking water in the affected regions. There are instances where such conditions prevailed over the state Droughts and famines affect the agriculture production to severe extent, create problem of food grain, fodder and water to the critical level of survival. Socio-economic conditions of the people is dwindled as agriculture is their source of livelihood and all expectations of the people depend on the ensuing crop. The immediate problems they face are availability of food grains, employment for the duration of next monsoon. Number of livestock is more than human population and problem of their survival becomes acute. Delay in relief measures create famine conditions.

The outcome of this research study is the fact that droughts occur with monsoon failure is not only impact of natural factors but human interface is equally responsible which has disturbed monsoon cycle to great extent. Uncontrolled felling of trees, clearance of vegetation for cultivation purposes, degradation of natural fertility by chemicals, ground water depletion much in excess than charging level and dryness in the atmosphere are all responsible. The environment of the state is seriously degraded and occurrence of droughts have affected socio-economic conditions of the cultivators and others directly and indirectly associated with agriculture.

Drought proofing activities initiated for scaling down the impact of droughts could not yield desired results in view of faulty process of implementation in isolation without associating the people in decision making process. The problems cropping up as impact of drought are increasing continuously. For minimizing the impact of drought, it is necessary to focus the activities on rain water harvesting, plantation activities in 33 percent of area comprising of degraded forests, waste land, barren land and permanent pastures is necessary to treat for existing deficiencies, This area is to be used under plantation activities under timber trees, horticulture trees and trees able to meet fuel wood and fodder requirements. In addition the gap area is to be used for pasture development.

For increasing the cultivated area, fallow and marginal lands need to be treated to meet existing deficiencies. Total cultivation practices can be possible in 60 percent of state geographical area. Presently the fallow lands are ranging between 40-50 percent of cultivable area is not a genuine practice for meeting the agriculture requirement of the people. 33 percent vegetated area would be helpful in attracting monsoon to the state and also retention of moisture in cultivated lands. The plantation activities would help in creation of additional employment generation to the tune of 80 lakhs to one crore and poverty alleviation would be possible.

The occurrence of drought due to rainfall failure would continue but its impact and magnitude would be decreased significantly as people of the affected regions would have sustainable livelihood support from plantation and livestock rearing. The socio-economic conditions of the people of the state would change to significant extent The environment and ecology of the state would be regained to most vital extent and Rajasthan state would become one of the most vulnerable state of the country.

Jaipur **T.S. Chouhan**

Acknowledgement

I express my humble gratitude towards Late Prof. J. K. Jain who help me at all stages to prepared this book. He spared his valuable time to make me understand multifaceted aspects of drought. He always encouraged me to proceed ahead.

The present book outcome of hard work and visits of various agro-climatic zones of the state to understand their conditions and situations specially during the disastrous conditions of drought. This work is original contribution based in the geo-physical conditions of the state. I have tried to remain impartial in total narration of factual conditions of the state.

T.S. Chouhan

Contents

Contents

1

INTRODUCTION

Droughts are one of the most serious natural hazards to mankind. They have been the root causes of famines in many parts of the world and more so in our country where agricultural success and prosperity are virtually dependent on the amount and distribution of rainfall. Droughts are usually defined as periods of dryness due to lack of rain though the concept varies from place to place depending upon the normal climatic conditions, available water resources, land-use, agricultural practices and various other economic activities of the region. In magnitude terms, drought is a relative term, for what is droughty for one region may be quite normal or even favourable to another which, therefore, leads to the assessment of drought and its effects not in terms of rainfall but only through water balance criteria. Droughts need not be and in fact are not continuous phenomena but vary in intensity, duration as well as frequency and thus dryness of the same magnitude may not necessarily produce drought of the same severity even at the same location whose economic activity and the land-use strategy may vary from time to time. It is, therefore, eminent that droughts arc natural occurrence, their impact can be mitigated and their adverse effects can be minimized, provided there is advance knowledge about their incidence and the people are technologically advanced to combat those. In this respect, drought climatology opened up a new field of research investigations into the several aspects of droughts, their origin, nature, frequencies, effects etc.

Since ages, the man has been living in balance with the natural environment of an arid desert eco-system. Though characterized by low and extremely uncertain rainfall resulting in harsh climatic conditions, yet the area is not devoid of life support systems. For the past thousands of years, the Rajasthan part of Indian Desert has been an unique human eco-system, all life forms- human, animal and vegetation have survived by evolving delicate and precarious relationship with the fragile eco-system. This has meant working out a symbolic relationship between people and the environment. The most marvelous inheritance of this civilization is its great physio-cultural institutions and bio-technological capacities to sustain a variety of life forms. This region is compared to any other in the world, has the highest biotic concentration of human and livestock population. Over time, a unique socio-ecological organism, a multi-ethnic, multi-caste hierarchical society interwoven in a symbiotic relationship, was born.

Now-a-days the general problem of Indian arid, semi arid and sub-humid environment is essentially one of human ecology. The man has one of the most important and active agents of the biotic interference in the fragile eco-system of arid zone of Rajasthan. Over-exploitation of natural resources by man has disturbed the delicate ecological balance. Over-grazed lands, shrinking forests and eroded agricultural fields show the imprints of man's activities on his environment. The way of life of the people in the region and their sources of livelihood have chiefly been conditioned by disturbances in ecological balance due to the severity of the arid, semi-arid and sub-humid climate. This has led to the degradation of vegetation and diminution of crop potential decreasing productivity per unit area. These characteristics of desertification have human significance. Moreover, while deserts are natural phenomena, desertification is human process. This poses a potential threat to resources and ultimately the lives and livelihood of the inhabitants.

Rajasthan with its diverse climates and especially because of its strong dependence of monsoon rainfall for its agricultural and hydrological activities experienced in the past several famines of both widespread and local nature. Though every severe drought can strike an unprepared community and cripple its economy through famine, not all famines are due to drought since several other natural disasters like floods, earth-quakes, epidemics etc. can produce shortage or contamination of available food and water and lead to famine conditions. There were certain major droughts affecting total state several times, the main reason for famine is failure of monsoon. It is also fact that western Rajasthan had faced more famines in comparison to the eastern part of the state.

Droughts have long been recognized as one of the serious catastrophies of nature against man, no systematic methods have yet been developed either for their complete understanding for their prediction. Non-existence of reliable periodicities in the occurrence of droughts reported by investigators from different parts of the world is responsible for obscurity still lying in the prediction of these phenomena. It appears thus that there is no easy way of tackling this problem except through dynamical and synoptic treatment of the atmospheric circulation systems Acute water shortages arising from lack of sufficient rain being the primary cause, drought form one extreme end of the hydrologic while floods are at the other end.

A clear understanding and better appreciation of the problem of droughts and aridity become possible after the subject of climatic classification came to be developed at the turn of the century. Among the many systems and schemes that were put forward from time to time, that of Thornthwaite (1948) has gained world-wide acceptance and recognition, for it is based on the rational water balance approach which itself has been developed from the universal hydrologic cycle.

According to this scheme. There are five major groups of climates namely, per-humid, humid, sub-humid, moist sub-humid and dry sub-humid, semi-arid and arid based upon a parameter called Moisture Index (Im) derived from water surplus and water deficit obtainable from book-keeping procedure for the computation of water balance. The per humid and the humid climatic groups, as they to do the humid category, are naturally drought tolerant while the two dry climates, the semi-arid and the arid have great drought proneness; permanent occurrence or existence of droughts in arid and semi-arid climates is responsible for the formation of desertic conditions there. Sub-humid category of climates posses critical water balances that functions violently from year to year and are, therefore, liable to incidence of droughts of various kinds. Contingent droughts occur almost anywhere but are the most characteristic of sub-humid climates. Though they are usually short duration and are confined to limited areas, they are the most treacherous in view of their great intensity and critical times of incidence i.e., when there is maximum need for water.

According to water balance concept, drought is a physical condition of the environment, in which the amount of water available from precipitation and soil reservoir is not sufficient to meet the demands of evapotranspiration. Employing the book-keeping procedure, it is possible to compare the water supply by precipitation with water needs through potential evapotranspiration using soil as a moisture reservoir and compute the water surplus, water deficiency and the actual evapotranspiration on a monthly, seasonal or even daily basis. In drought climatology, it is the water deficiency that is important and expressing it as a percentage ratio of the water needs, one can obtain the Aridity Index (Ia), which is a very convenient parameter for studying the various aspects of droughts. Categorization of droughts can be made in the following manner

	Departure of Ia from the median	**Drought Category**
(i)	Less than ½ S	Moderate
(ii)	Between S ½ and S	Large
(iii)	Between S and 2 S	Severe
(iv)	Greater than 2 S	Disastrous

Where S is the standard deviation of Aridity Index during the period of investigation.

1.1 STATEMENT OF THE PROBLEM

Problem of drought is significantly threatening in Rajasthan as the cycle of drought is once in two years barring the intensity and area affected. There are occasions when the total state remained gripped with the famine for one or consecutive two years. The people of the state still remember Chappania Akal for devastating loss of human beings and livestock. Periodical failure of rains and consequent droughts leading to famine is the recurring feature of the area. Changes in the resource use and socio-economic parameters take place as a result of these phenomena. Large scale socio-economic distress causing deviation of the normal patterns of behaviour involving changes in the social and economic values among the population to cope with the ill effects of drought take place in the area. Depopulation of the villages, disruption of social institutions and relationships, disturbances in agrarian sector including increasingly subsistence oriented cropping pattern and livestock are the recurring consequences of droughts.

Rajasthan state possesses about 10.40 percent of the geographical area of the country, 5.50 percent of the country population and about one percent of the water resources. The state economy is mainly dependent on agricultural practices, which is mostly monsoon based. There are very few perennial rivers systems and some part of the state has canal irrigation facilities. Most of the cultivated land is single cropped, based on the mercy of monsoon, which is erratic and uncertain. In view of scanty rains, surface water harvesting is very meagre and most of the part of the state is irrigated through ground water resources, which is at the significant depth and recharging level of ground water is quite low than the requirement causing significant problems of water depletion. Most of the state is facing serious ground water problems and categorized in to dark zone due to over-exploitation of ground water. The development process of the state is moving forward within these limitations.

Western part of the state receives very low rainfall ranging between 200 to 300 mm and uncertainty of the monsoon creates various problems like late arrival of monsoon, timely arrival of monsoon but failure at the later stage and early withdrawal causing serious problems in cultivation practices. Even the limited rainfall with timely arrival and its continuance with regular intervals can provide bumper crops, as it happens once in five years. In the monsoon cycle of five years, at least two years are drought years with moderate to very severe intensity. This dwindles the total developmental process of the state and people suffer a lot in view of such problems.

The annual predictions of monsoon are still not accurate to the desired level as they provide general situation of the country as a whole. India is very vast country, where such predictions do not reflect the regional issues. Generally the overall monsoon trend varies to large extent in terms of arrival and departure of the monsoon in the country. Rajasthan receives monsoon from both the systems i.e., Arabian Sea and Bay of Bengal but its arrival is unable to be predicted even before one week. The satellite position of clouds starting from coastal areas change the position quite frequently and all the predictions made for their approach areas becomes false. Therefore, the analysis part of the monsoon is quite weak and unreliable. The cultivators taking initiatives on the basis of the forecasts suffer significantly in view of changing wind directions.

In addition to the predictions, the occurrence of rainfall is varied in area and frequency. The remote sensed imageries and analysis thereof could not distinguish in between the high residual cloud cover or by thin clouds unidentified by visible band cloud masking or due to off-nadir viewing as a result of insufficient cloud free passes in the compositing period, require validation before use in drought assessment. Heavy rainfall at the time of flowering/ heading can disturb grain development and lead to yield reduction, which however may remain undetected by the satellite, which shows healthy canopy conditions.

During the total process of cultivation of monsoon crop from sowing to harvesting, rainfall has very crucial and significant role for productivity yield and in success or failure of crops. Every spell of crop cycle is critical from rainfall point of view and at each spell, cultivators need to adopt various safety and precautionary measures, but the failure of rainfall completely dwindles the crop cycle. Therefore, chances of crop damage are valid at each spell and the final outcome in terms of crop is the ultimate result of various factors played its role in the crop cycle duration. In true terms, the meteorologists, geographers and all other specialists responsible for prediction of the monsoon are simply the facilitators to the cultivators of the country or the state, but the ultimate sufferer is the cultivator only who has every thing at stake. Therefore, it is a naked truth that the cultivator of fifty years back or of present time has same situations and no significant change is visualized in their condition.

Drought leads to moisture stress, which, in turn, affects crop production adversely. Efforts have been made to devise means of stabilizing dryland agriculture by evolving contingent crop production strategy in different whether conditions. The results so far achieved are choice of suitable crops in view of rainfall conditions, selection of crop varieties, alternative crop strategies, mid-season corrective measures and crop-life saving devices. Often times, certain crops are grown more for convenience or by convention. This is because of the first priority of a cultivator continues to be production of food for his family and feed for his livestock. On the other hand, selection of crop suiting the environment leads to increased and stabilized production.

1.2 METEOROLOGICAL DROUGHT

Meteorological drought is mainly based on the degree of dryness and its duration in days due to deficient rainfall over a region for sufficiently long period, more than a particular number of days with precipitation less than specified small amount, a sustained period of time without significant rainfall, deficient rainfall below reference value, a threshold rainfall value to distinguish drought occurrence, seasonal rainfall deficiency by 25 percent or more, the dryness to the normal value of rainfall. All these factors reflect over the localized areas in terms of vegetation losses to relative extent. These are highly region oriented because of prevailing meteorological or climatological conditions of the area. Some definitions include other atmospheric parameters also besides rainfall like wet bulb temperatures, humidity and temperatures, vapour pressure deficiency, period of strong winds with low precipitation, high temperatures and low relative humidity.

1.3 HYDROLOGICAL DROUGHT

The hydrological drought consider dry spells of different intensity and duration which affect the surface or sub-surface water table as well as river and stream flows and level of reservoirs. Some defined it as a period during which stream flows are inadequate to supply sufficient water for its management during the year, when the aggregate is less than the long term average. The hydrological affect much over the catchment areas of the streams and the moisture in the surrounding region is decreased considerably. Furthermore, the underground water charge is seriously affected and in most of the cases, the ground water sources become dry or possess very limited water charge not sufficient to draw for half an hour. This situation completely eliminates the prospect of irrigated crops, which cultivator generally hopes to get after failure of monsoon.

In Rajasthan, where most of the area is dependent on ground water resources for cultivation as well as for drinking water, the agriculture is suffered significantly as the first charge becomes to provide drinking water to cities and villages for survival of the people and livestock. In the situation of hydrological drought, the reservoirs and under ground water sources are focused to meet the drinking water mainly to urban population and there are severe clashes between the people residing near water points. In all the circumstances, the cultivator is the main sufferer. In various parts of the state, the general ground water depletion level has reached up to one meter per year, which is a matter of general concerns, but in case of drought conditions, the administration feels utmost necessary to survive the people by providing drinking water from any possible measure.

1.4 AGRICULTURAL DROUGHT

The agricultural drought is highly related to the meteorological drought because four-fifth of the cropped area of Rajasthan is rain-fed. The main problem of the rain-fed area is to minimize the loss of agricultural production by sufficient environmental management in drought years. Any definition of agricultural drought has to be associated with the rainfall and its variations. Amount of rainfall and its derived parameters are correlated with agricultural output over an area. This gives a preliminary view of impact of deficient rainfall on soil moisture and thereby agricultural drought. Modifications are reported to correlate agricultural drought with rainfall distribution. Rainfall less than first decile, four consecutive dry weeks during kharif season or six such weeks over the rest of the year and rainfall less than a threshold value for different crops over an area. These are associated with some drawbacks as the rainfall is an average over a non-homogeneous area and is for longer periods of time like a season or so. However this is beginning to quantify the agricultural drought. To overcome some of the defects, indices of rainfall on short term basis were proposed and the cumulative index was correlated with the final yields. This is used for demarcation of drought prone areas too.

The availability of sufficient soil moisture to the roots of the plant at different layers of soil is an important parameter for healthy growth. Any shortfall of soil moisture to meet fully the demand for evapo-transpiration lower than the potential, hampers the growth, thereby reduces the final yield. The soil moisture concept is used to study the agricultural drought based on water balance technique suggested by Thronthwaite, taking into account the various processes that are involved in hydrological cycle. This technique gave a method of computing the available soil moisture at the root zone of the crop either by a single or multi-layer soil profiles.

1.5 PARAMETERS FOR ASSESSMENT OF DROUGHT INTENSITY

On the failure of monsoon and possible crop damage, the assessment of drought intensity is made in terms of crop loss and problem of drinking water. The exercise of crop damage is started in the month of September - October on the basis of crop damage and carried out by the district administration through the revenue agency. For this purpose Disaster Management and Relief Department is the administrative department, activated for assessment of crop damage in the state. The District Collector is responsible for the assessment of the drought situation in the respective district and the revenue agency of each district is put for assessment of the impact of drought. The parameters for assessment of drought intensity are combination of various issues like crop damage, problem of drinking water etc.

The assessment of crop damage is done by field level functionaries of the revenue agency, where assessment is made in respect of each villages through the visual assessment of crop damage up to 25 percent, between 25-50 percent, between 50 to 75 percent and between 75 to 100 percent. The first category of crop damage up to 25 percent is treated as the normal phenomena. The crop damage up to 50 percent is treated as moderate drought and treated that the cultivators are able to attain their livelihood with the forthcoming crop produce and fodder. The crop damage up to 75 percent is treated as severe drought conditions and it is deemed that the cultivator is significantly depressed with the situation and requires some livelihood support for the remaining part of the year. The situation where the crop damage is cent-percent, the situation is treated as the most devastating and deemed it as very unfortunate situation for the people of the area mostly of the cultivators and agricultural labour, who are dependent on the agriculture activities for their livelihood.

The impact of the drought is also assessed in terms of drinking water problem in view of drought conditions. The area specific position is segregated in to blocks and later for the district as a whole. In case of crop damage up to 75 percent and up to 100 percent, a memorandum is prepared by the District Collector indicating the impact of drought in terms of severity in the district and suggests possible measures to face the problem in the ensuing period in the district. The district memorandum contains the details relating to the waiver land revenue or its deferment in view of the severity of the drought, postponement of recovery of agricultural loans due from cultivators. In addition the measure to assist the people affected from drought is also incorporated in the memorandum citing the necessity of starting the wage activities to help the cultivators and other affected people quantifying their possible number. The situation of drinking water is also analyzed in the document and suggestions for providing water in hard hit areas of the district is also done. These memorandum are submitted to the state government where the district-wise situation is analyzed and state memorandum is prepared for submission to the central government indicating the area affected and requirement of funds to meet till the next pre-monsoon duration.

The central government on receipt of the state memorandum analyses the state situation and a team of experts is sent to state to have first hand review of the situation and recommends the Government of India about their assessment and recommends the requisite funds for the drought affected areas of the state. With this process, the famine funds are generated and the management of drought proofing measures are carried out. The most recent drought was of 2002, when all the districts of the state were declared drought affected. The relief measures are started after receipt of the funds from the central government. The State and Central Governments are supposed to have Natural Calamity Relief fund at their level. In case of the low intensity or very little affected area, the state government is supposed to start measures from the funds available with them. In the event of severe intensity and coverage of larger area the central funds are made available to the state to overcome the situation.

Since the state is affected by drought situations, the aforesaid process is the regular phenomena and every state government tries to receive maximum funds from the central government stressing the problem at various levels. Rajasthan Government is not an exception to pose the aggravated drought situation to the maximum possible extent. The drought has become a crucial factor between the central and the state government as per centre-state relations.

1.6 DROUGHT MANAGEMENT

In the event of drought situations, the position of the affected areas is regularly assessed at the apex level in the state government and district-wise ceiling of labour engagement in relief works is drawn for each duration on fortnightly basis, which is reviewed from time to time and the ceiling is increased or decreased in view of requirement. The ultimate objective of the relief measures is to provide the support to the people through wage oriented activities to attain their livelihood. The payment of the wages is done in terms of food-grain and cash to help the people to restore their purchasing power and survive in the hard-hit conditions. In view of regular droughts the state government prefers to engage the labour for creation of assets of permanent nature.

Water supply arrangements are made through tankers and in case of severity of the situation, assistance of the railways is also sought to meet the requirement. Food and water are the essential items for every human beings and state government is required to take every precaution to save the people form hunger and thirst. Other measures of drought management are supply of fodder to the affected areas and opening of camps for upkeep of the livestock, which the owners are unable to manage due to adverse situation.

Droughts are the natural calamities and hamper the process of development. The environment of the affected areas is disturbed in case of severe conditions. The people are compelled to move to moneylenders for meeting their emergent requirements like diseases or social obligations. The droughts affect the people at large but the ultimate sufferers are the poor, daily wage earners. Since the state economy is based on agriculture the total economy of the state suffer from the problem. Even the business community is severely affected due to decrease purchasing power of the rural people. The main cause of the backwardness of the state is the recurrent droughts where no permanent solution is insight in the near future.

2.0 REVIEW OF LITERATURE

Since the early sixties of the last century, when the country suffered from severe droughts due to erratic monsoon behaviour, climatologists have been devoting much attention to drought climatology, especially in the context of monsoon failures. The problem of droughts has been examined from different angles, particularly in the context of deficient rainfall during the monsoon season either on account of late onset or early withdrawal or both, irregular and inadequate distribution of rainfall during the season. Bhalme and Mooley (1979) studied large-scale droughts in relation to monsoon circulation over the Indian region having been modified or impeded due to several extraneous causes outside the tropical latitudes. Mooley, Parthasarathy and Munot (1981) have reported a study of large-scale droughts in India and their impact on economy, which should be immense value to applied geographers in tackling water scarcity situations. The general paper of Mooley, Parthasarathy, Sontakke and Munot (1981) about the annual rain water over India and its impact on national economy is quite relevant in this connection. The works of Gupta (1981) for Rajasthan, Swadesh Mishra (1983) for Gangetic West Bengal, Biswas and Basonkar (1982) for Gujarat, Roy and Mittal (1982) for Delhi and Tali and Qureshi (1979) for southern Rajasthan are also about rainfall fluctuations and failures for smaller areas but having the same objective of the study of droughts especially in the agricultural context. Subramanyam and Vishwanadham (1979) too discuss the rainfall patterns of the Visakhapatnam district from an agricultural angle.

As in early times, water balance climatology stands out prominently judged from the number of publications and the variety of topics covered. Many of the papers are of course concerned with the concepts of droughts (Subrahmanyam 1981)and different aspects of drought climatology of individual regions, Ram Mohan and Subrahmanyam (1980) about the semi arid region of Tamilnadu; Ratnam, Hosmani and Joshi (1981), Ratnam and Joshi (1983) pertaining to Karnataka, Subramaniam and Sambasiva Rao (1982) for Maharashtra; Subramaniam and Venkateswara Rao (1983) about Rayalaseema in Andhra Pradesh and Subramaniam and Vinayak (1982) for Trivendrum in Kerala. Ram Mohan and Subrahmayam (1983) discuss the water balance criteria in relation to agriculture in general, while Sambasiva Rao and Kalavathy (1983) deal with water balance and drought categories for agricultural operations in Madurai district from the point of view of cropping patterns. Certain aspects of aridity, spread and severity of droughts in Andhra Pradesh were treated by Hemamalini (1981) in her paper. Jutta Dikshit

(1980) studied and reported identification of drought prone areas in Maharashtra and explored the possibilities of crop expectations in different seasons.

On the general topic of classification of climates, there were three major papers during the period; Subrahmanyam and Ram Mohan (1980) on a scheme of classification of monsoon climates and stability of their moisture regime employing for the first time an Index of moisture availability; Subramaniam and Prasada Rao (1981) on the eco-climates of the Rajasthan State and the paper of Subramaniam and Sambasive Rao (1984) on the climatic classification of Maharashtra from an agricultural point of view.

Application of water balance concepts for water potential estimation and water resource development is the subject matter of two papers by Subrahmanuam, Ram Mohan and Pardhasardhi (1979) and Subrahmanyam (1983) the first dealing with the appraisal of ground water potential for the dry region South India and the second on the hydro-climatology and water balance of the Chandrampalem basin near Visakhapatnam. Sastry, Ramakrishna and Bhandari (1982) apply the same concepts for the study of drought climatology of Gerhiya catchment in Rajasthan. Subrahmanyam and Sharma (1980) in their paper on the evolution of Indian climates with regard to geographical location, mountain systems and monsoonal circulation, discuss the various climatic categories of the Indian region and water use planning, an aspect of extreme value in applied geography. Subrahmanyam's edited book (1983) on applied climatology has several articles dealing with the concepts and use of water balance in agricultural and hydrological planning and operations that should stimulate the interest of all geographers concerned with water resource development.

Nor'westers are severe thunderstorms during the hot weather season in the north-eastern Indian region about which chowdhury and Majumdar (1983) published some work, especially with regard to the synoptic climatology of these phenomena. In a similar manner the paper on hailstorms by Chowdhury and Banerjee (1983) is a very informative one since both of them are well known for their sudden occurrence and destructive nature.

Probable relations between weather and solar activity are an age-old problem on which not much work has been done or reported; the available literature again is scanty and inconclusive. Yet, there are some contributions; Reddy, Mukherjee and Ramanamurty (1976), Reddy and Ramanamurty (1978) and Bhalme, Reddy, Mooley and Ramanamurty 91981) reporting studies on solar relations to weather in general and to rainfall in particular. These may be examined further by geographical climatologists by pooling up the scattered data and available information for more affirmative findings.

A new impetus to this field of applied climatology has been provided by the scientists of the Indian Institute of Tropical Meteorology. Especially by Pant (1979) on tree ring analysis and earth's past climates, in which paper the field of study has been described in general and palaeo-climatological significance and potentialities as far as the Indian region is concerned, in particular. In two other papers (Pant

1983; Bilham, Pant and Jacoby 1983), the work has been implied with special reference to certain selected tree species in India. This is very fascinating field of climatology, which may be taken up by geographers with some background of geology and forestry, not only for reconstructing the links in the dendro-climatology of the region but even for projecting the same into the future in the light of the currently available evidences.

Micro-climatological research was mainly on application of the physical principles to cropped environments. Ramakrishna, Singh and Singh (1982) report the influence of systems of planting pearl millet on generation and maintenance of crop climates. On the other hand, human coi-climatology stands out very prominently on account of the growing interest in weather with regard to human health and comfort. At an international symposium in Madras, Subrahmanyam and Sharma (1981) presented a paper on the human comfort zones in south India, employing the thermal efficiency concept of Thornthwaite which was not attempted by any other scholar in the past. Chowdhury and Genesan (1981) studied the physio-climatic regimes of India with the same objective. Rama Rao (1977) has presented a new approach of using basic meteorological factors for computing discomfort indices, while Prasad and Pawar (1982) describe the uncomfortable conditions over Bombay during the pre and post monsoon months as a matter of general information.

A field of meso/micro-climatology that gained utmost importance during recent years is climatology of urban areas for the proper planning and construction of city buildings and other structures, particularly in the context of growing environmental problems of a climatological nature. Though a considerable number of papers have appeared on these studies, the paper of Subrahmanyam, Sachi Devi and Viswandham (1982) presents a detailed climatological study of Visakhapatnam from a thermal point of view. An extension of the same work by Surbahmanyam, Sastry and Sachi Devi (1978) attempts a comparison of the temperature and humidity characteristics of urban and non-urban areas around Visakhapatnam and discusses the peculiarities of urban and rural climatology as determined by structures, layouts, building materials as well as population numbers and densities.

High erosion and sediment transport occur widely in many parts of the Indian sub-continent. Although the sediment yields are influenced by several physical and climatic factors. At the same time, forest destruction and changing land use in recent decades have also contributed in increasing the sediment production. However, very limited information is available regarding erosion rates from Indian river's catchment areas, to evaluate the role of deforestation and agricultural expansion on sediment yield. In this regard, several studies carried out by Rawat and his co-workers are important (Tawat *et al.* 1996 and Rawat and Rai, 1997). Measurements of run-off and erosion rates were carried out on five micro-watersheds over a period of three years, in the Khulgad Basin in the Kumaun Himalaya (Rawat and Rai, 1997). The average rate of erosion in the area was estimated to be 0.07 mm/yr or 186 t/sq.km/y. the rate of erosion is about 5< times higher in the anthropogenically disturbed areas and the figure increases to 10 times for the tectonically disturbed lands. the adverse impact of human activities on soil erosion was also demonstrated in another study by Singh and Dubey (1996). The study was designed to estimate the loss of soil nutrients from gully basin of Deoghat area. The results of the study indicate that the amount of loss from human impacted gullies is of higher order than from the undisturbed gullies. In another paper, Singh and Dubey (1996) have shown that the intensity of rainfall in more significant climatic parameter than the total amount of rainfall for gully erosion. The rates of erosion and badland development in the lateritic terrain in Santiniketan were studied by Das and Bandyopadhyay (1996). They found higher rates in the non-laterite capped areas than the laterite capped areas. Basu and De (1998) taking in to consideration some variables of the universal soil loss equation, have attempted to identify zones of soil erosion in a small basin in Darjeeling Himalaya.

Weathering profiles and regolith are one of the most strinking features in many parts of India. Although, these formations have been described for a long time, very few attempts have been made to actually ascertain the rate of weathering profile development. Gole (1997) attempted to establish the rate of weathering profile development and the rate of denudation in a small watershed in Koyana basin. On the basis of the assessment of the thickness of weathering profiles, hydro-chemical analysis and assessment of silica losses, the estimated age of regolith formation and denudation was found to be between 20ka and 30ka.

Colluvial deposits, resulting from mass movement and fluvial erosion of regolith on hill-slopes are quite widespread and occur in many parts of India. A detailed study of the culluvial deposits in north-west Deccan was carried out by Joshi and Kale (1997). The characteristics of the colluvial deposits at four sites were evaluated on the basis of textural, geo-chemical, x-ray diffraction and scanning electron microscope data. The data were used to understand the weathering characteristics and nature of hill slope processes during the late quaternary period. On the basis of this study, the authors have identified three phases in the evolution of colluvium.

Use and application of remote sensing techniques have been focus of some studies in recent years. Anbarasu and Rajamanickam (1997) carried out a study of abandoned channels flowing over the coastal plains of northern Tamilnadu. Examination and interpretation of remote sensing data suggest that in the area some rivers are migrating northward, whilst other are shifting southward. The shifting of the river courses has been explained in terms of neo-tectonic activity in the area. Likewise, in the last few decades, several studies have been carried out to identify palaeo-channels and former river courses, particularly of Saraswati river in the Thar Desert, on the basis of satellite images. In a recent study, Kar (1999) has interpreted the radar imagery of south-eastern Thar to identify palaeo-valleys of Luni river in the alluvial plains between Jodhpur and Pali. The results of the study indicate that the present course of the Luni river between Kankani, tilwara and the confluence with the Jawai is a relatively recent development.

Geomorphological mapping of Assam state was carried out by Goswami *et al.* (1996), with the help of remote sensing techniques. Different types of depositional, denudational and structural landform assemblages were identified and mapped. In the desert areas, it is often difficult to carry out extensive fieldwork and undertake mapping. Therefore, Tsunekawa *et al.* (1996) used NOAA/ AVHRR satellite data to classify land cover for Rajasthan. On the basis of normalized difference vegetation index, nine land cover classes were identified and mapped. The study by Goswamy *et al.* (1996) for the state of Assam prove the immense utility of remote sensing techniques for geomorphic and land use mapping of large areas.

In a micro-level study, Venugopal (1998) used remote sensed data to monitor the infestation of water hyacinths in Bangalore city. Using SPOT multidata images and NDVI, new infestation in fresh water tanks was observed. Global positioning system is a new tool for accurately measuring the coordinates and elevation of any point, particularly in less accessible terrains. Although, earth scientists abroad are using the system, its utility in India has not yet been recognized. An exercise to test the utility of GPS for measuring sand dunes was carried out by Kar *et al.* (1998). The data were used by the author to prepare a digital elevation model. The case study proves the usefulness of the instrument for field studies in inaccessible areas.

The papers of climatology are generally based on the spatial and temporal analysis of meteorological parameters. Rainfall is usually the most important and, therefore, it is not surprising that a large number of existing studies deal with some of the aspects of rainfall in Upper Assam. The study shows the effect of complex physiography of the region on the distribution of rainfall. The results of the analysis suggest that while the month of July record maximum rainfall, two secondary peaks are also observed either in April or September or in both the months.

The rain spells over Pune city, with respect to duration, yield, intensity and their contribution to seasonal rainfall were investigated by Deosthali (1997). The study indicates that more than half of the seasonal rainfall is contributed by rain spells of > 3 hours duration and most intense spells of 40 mm/ hour occur in July and October. In another study of the same city, Gadgil and Patil (1998) examined and analyzed the synoptic conditions associated with heavy rainfall. Between 1901 and 1988 heavy rainfall and very heavy rainfall was recorded on 67 and 5 occasions during the period 1979 and 1988 were analyzed in detail by the authors. The analysis suggests that on sixty percent of the occasions, heavy rainfall was associated with the persistence of depressions or low pressure systems over both the adjoining seas. The heavy rainfall at Pune was found to be the combined result of several synoptic situations. In an applied study, Deosthali (1998) analyzed daily rainfall data for 41 stations to evaluate the sustainability of sorghum (jowar) in the scarcity zone of western Maharashtra. On the basis of the analysis, the author has concluded that the scarcity zone is more suitable for rabi sorghum rather than kharif sorghum.

Next to rainfall, ambient air temperature is an important weather parameter from the stand point of biodata and agriculture. Certain crops like grapes are seriously affected in the temperature falls below freezing point. An attempt was made by Deosthali and Payyappalli (1997) to analyze the characteristics of cold waves and predict daily minimum temperature of Nasik region in Maharashtra, known for quality grapes. On the basis of daily data of various weather parameters for 25 years and synoptic charts of upper atmosphere, the authors have identified 12 instances of severe cold waves. A regression model was also developed that uses previous day's weather data to predict minimum temperature with 74 percent accuracy. Apart from minimum and maximum temperature, the information about diurnal variation in temperature is also critical for many climatological and agricultural purposes. For constructing hourly temperature variations, a statistical model was also developed by Deosthali and Iyer (1996) using cosine functions. The model was found to be useful for monitoring growth of rainfed crops and to predict crop yields.

The Deccan Geographer published a special issue on climatology in June, 1997. The issue includes several papers by scientists from the Indian Meteorology Department and Indian Institute of Tropical Meteorology as well as some geographers. One of the papers in this issue is by Hema Malini (1997), which tries to evaluate the probable effect of global warming on thermal characteristics and moisture regime conditions of India. Using water balance methods applied to Hokkaido, Japan and India, she had arrived at the conclusion that global warming may have lesser effect on tropical areas in terms of hydrological conditions and climatic regimes than the middle latitudes. She had also concluded that there is likely to be negligible change in the thermal conditions of India, but a shift in moisture regime is expected to be higher.

An assessment of the impact of urbanization on the climate of Pune city was carried out by Deosthali (1999) by applying the bio-climatic index. The results of the analysis indicate that in the mega-city of Pune, the temperature moisture conditions between 0830 and 1730 hours are not comfortable for nearly two third time of the year, that is from March to November. The hydro-meteorological conditions in the Himalayas were analyzed by Hofer (1998) to understand the causes of flooding in the foothills. On the basis of the analysis, the author concluded that the floods in the lowlands of the Ganga and Brahamaputra have not increased and that the land use changes in the Himalayas have no connection with recent floods in the plains. The findings are contrary to the views expressed by many environmentalists in recent years.

In addition to these studies, there are few more papers on climatology in various Indian journals, but they have not been included in this review because they have been authored by non-geographers. In brief, the research done by geographers shows that while the number of case studies on various topics has increased over years, our knowledge regarding the physical geography of the Indian sub-continent remains incomplete to large extent.

Since climate has its all-pervading impact on the growth, development and distribution of the biotic as well as abiotic features on the earth, climatology stands in the front line of studies in physical geography. But in proportion to its potential significance and relevance, the study of climate, in fact, find no due weightage in the work of geographers. Some branches of physical geography, especially bio-morphology, whose documentation is traditionally bulky, still in the process of being enriched rapidly with new literature. In contrary to this, geographers' work on climatology are generally very limited and it is more so in case of Indian geographers. A very few Indian geographers are found to concentrate on the study of climatology and thereby this branch presently appears to be ceased to be a lively field of study and research.

Climate has been subject to changes on all times and space-scales. Climatic changes being transitional threaten to affect the whole planet. In India, in view of the increasing concentration of greenhouse gases in the atmosphere, many scientists suggest that there is likelihood of perceptible climate change. Singh (2001) remarks that climate change direction in India is still not unequivocal and some controversies persist particularly about the sources of its change, magnitude, trends, spatial extent, natural and socio-economic impact etc. While analyzing climate variability and change in India, Singh's paper concludes that observational studies on the climate change in the country as yet do not throw any certain picture with regard to trends based on observational data for the last 100 years or so. The paper further suggests that there is the need to investigate the issue of climate change and variability with a good data base comprising of more relevant atmospheric data and derive precise conclusions through different observational studies and simulation models.

In presidential address delivered at the XIV conference of the Indian Institute of Geo-morphologists held at the University of Annamalai, Chidambaram, Sharma (2002) emphasized on the fact that the increasing frequency of consecutive climate-related disasters has brought wide spread misery and huge economic losses to India. While discussing the state of India's climate amidst global climate change situation, he has shown that in the 20th century, the annual average temperature in India has increased by about 0.57* C. He also remarks that several global studies have indicated India is particularly vulnerable to climatic change, which will trigger a series of ecological and economic fall out in the country.

It is not surprising that the quest for global climatic phenomena often blinds us to what is going on around us. There lies the encouraging fact that some climatologists feel justified in concentrating upon the analysis, which gave greater insight into the climates that one experiences and can examine at the local and regional level. Deosthali (2000) lays emphasis on the need of knowledge of the agro-climatic conditions of an area for proper land management and his work thus deals with agro-climatic regionalization of Maharashtra using Reddy's agro-climatic analogue technique and thereby three main divisions viz. wet, wet-dry and dry have been delineated. Syiemlieh and Das (2003), while analyzing the trend of rainfall in the Brahmaputra Valley of Assam, find an increasing trend of rainfall in the valley. may be analyzed in the context of changes in the global circulation of the atmosphere, i.e., the moving in of the sub-tropical high to the north and the changes there in the Assam monsoon. Bora (2003) examines the monsoon controls in regard to the Brahmaputra floods in Assam and shows that the runoff pattern of Brahmaputra catchment in the Himalaya, which is largely responsible for occurrence of floods in the plains is primarily controlled by the quantity and natural distribution of monsoon precipitation in the area. He concludes that heavy monsoon rains have their direct bearing on the Brahmaputra floods and thus flooding in the Brahmaputra is truly monsoon-synchronized. Singh (2000-a) studied drought as a natural hazard in the sub-humid climatic region of India. Citing a case study of Palamau, he examines the incidence pattern of drought along with their impact as natural hazard in the study area. While analyzing the climate variability and its impact on crop productivity, Singh (2000-b) has clearly shown how crop productivity is affected by climate variability in Bihar Plateau. Singh's work sets a good example, how climatological knowledge finds suitable application in the field of agricultural geography. The work of Das Gupta (2002-b) also deals with agricultural meteorology and nicely discussed some tenets of ancient Indian agricultural meteorology.

Das Gupta (2002-a) explains in detail the concept of monsoon circulation held by the Vedic Aryans in his discussion about the growth and development of atmospheric science commenced in India during the early Vedic period and draws conclusion that despite lack of modern and sophisticated scientific instrumentation in the hoary past, the ancient Indian savants did some serious thinking about the process and science of weather phenomena prevailing in the sub-continent of India. He further observes that the Indian model of the hydrological cycle and the basic mechanism of the monsoon circulation identified by the Aryan scientists do not much deviate from the models adopted in the modern system of atmospheric science.

Indian monsoon is the life line of the country as the livelihood of the majority of its people depends upon timely and adequate monsoon rains. Indian monsoon is characterized by irregularities in its advent and retreat resulting in floods and droughts in one or the other part of the country. There may be some links between flood and drought years with the El-Nino episodes, which is considered as one of the sixteen parameters by the Indian Meteorological Department while forecasting the monsoon rainfall. The association between El-Nino and rainfall occurrence pattern has been examined by Ingle (2002) in his paper devoted to comparative analysis of El-Nino and rainfall over India and Goa during 1901-2000 which reveals that Goa being situated on the west coast of India, receives a fairly amount of monsoon rainfall, it has to face occasional droughts and floods. Further- more, the paper concludes that in Goa the El-Nino southern Oscillation years can be either drought or flood years. Hema Malini (2001) also shows the relationship between ENSO events and weather extremes through some analysis of India's climatic situation. In the same line, Singh (2000-c) has examined the incidence of drought and El-Nino events and concluded that there exists relationship between the both.

Cyclone occurrence along the Indian coast is a periodic phenomenon and its impact is evident both on the physical and human environment. The coastal region experiencing high incidence of cyclones lies in the northern part of Indian Ocean, which accounts for about 6 percent of the total global cyclones. Frequency of the storms is more over the Bay of Bengal affecting coastal areas of Orissa, West Bengal, Andhra Pradesh, Tamil Nadu than the states than the states located along the Arabian Sea. Cyclones and their impact have thus become an important theme in the climatological study in the Indian sub-continent. In this context, besides the studies conducted by the meteorological agencies, contribution of Indian geographers is noteworthy. Kaur (2000) has elaborately studied the cyclones occurring over Andhra coast in their historical perspectives, deals with the decadal incidence and district-wise incidence of cyclones over coastal Andhra Pradesh during 1891-1996 and month-wise and year-wise occurrence of the same during 1977-96 and finally examines their impact on both human and physical environment.

Regarding satellite studies, though a large number of papers have appeared on the meteorological aspects through remote-sensing techniques, the data accumulated during the period must be said to be not adequate enough for building up a climatological base. The main problem is the interpretation of satellite data, which require highest level of efficiency to derive the results. For want of proper specialization the accuracy level is affected severely and the inferences drawn are not correct to the requisite extent.

It may thus be seen that there has been a significant amount and quality of contribution to climatological research, which is of considerable interest to physical geography. However, it is important to note that most of it has come only from scientists, who have been meteorologically trained; the remaining few publications came from geographers, who again had advanced climatology in their university curriculum. There was not much rapport between geographers and meteorologists but even today, the situation is not too different primarily because the meteorologists consider themselves as a superior community concerned mainly with such theoretical aspects as fluid dynamics and numerical weather prediction. While the geographers are keenly interested and eagerly awaiting the meteorologists to cater to their requirements in various branches like agriculture development, weather resource management, urban planning, human health and disease etc, the meteorologists are not prepared to leave their ivory towers for serving the geographical community.

This impasse has to be broken and can be broken by imparting climatological knowledge to geographers at the university level by well designed and comprehensively planned curricula both in theory and practical work. There is climatology at an advanced level in Master's course in certain universities but this seems to be inadequate. More and more universities must come forward with such initiatives so that sufficient number of geographical climatologists could be trained to take up project oriented research investigations at the doctoral level in different branches of applied geography. The sooner this is done the better it is for the geography community in particular and for the country at large.

3.0 OBJECTIVES

With this background, the present study makes an attempt to examine various dimensions of drought in a regional specific context. The borad objectives of the study are as under:

(i) To analyze the impact of drought over the environment of the affected areas in terms of natural and human efforts.

(ii) To review the impact of the drought over the people of the affected areas over themselves, their cultivated lands and livestock.

(iii) To analyze the social and economic impact of drought over the area and the people.

(iv) Critically review the drought measures by the state during last five decades and their impact on natural resource system.

(v) Analysis the impact of short tem and long term measures adopted by the state government beneficial during the droughts in terms of its severity.

(vi) Whether the people of the state are accustomed to relief measures and start crying over the slightest impact of the drought as part of their habit.

(vii) To examine the procedure for declaring drought in the district is based on administrative or political basis.

(viii) To identify the method used by rural population to mitigate the effect of drought on savings, borrowings and asset liquidation.

(ix) To suggest instant, short and long term measures for drought proofing of chronically drought affected areas.

4.0 HYPOTHESIS

About the drought and famine in the state of Rajasthan, there are various hypothesis are prevalent, which are to be tested in the present research project. The main hypothesis about the proposed study are as follows:

1. There is significant impact of drought on family income, sources of livelihood, employment and consumption pattern especially dependent on agriculture directly or indirectly.
2. Droughts are occurring in view of human efforts related to unsystematic use of land, water and other natural resources
3. Every possible land is being put under cultivation in view of increasing population pressure. Accordingly the area under forests, pastures, common property resources are decreasing constantly.
4. Most of the state area is water deficient and the powerful people are using the available land and water resources for their own benefits.
5. Monsoon failure is regular phenomena and no requisite technology has been developed to forecast the drought conditions in various parts of the state. In case of early prediction, the impact of drought can be minimized.

6. Drought proofing measures could not yield the desired results and the severity of drought is increasing year after year. The watershed approach termed as most dependable measure to minimize the impact of drought has proved utter failure and no significant impact is visible in any part of the state.
7. Impact of drought is most in the arid and semi-arid regions of the state while other areas facing the drought are not affected to the similar extent. This is also true for the irrigated areas of the state.

5.0 METHODOLOGY

For the research study the information has been collected from the secondary sources from various departments like Disaster Management Department of the State Government (Relief Department), Central Arid Zone Research institute, Jodhpur, Central Ground Water Board, State Ground Water Department, Indian Meteorological Department, District Offices, Irrigation Department for creation of data base.

The satellite data analysis has been obtained from State Remote Sensing Application Centre, Jodhpur and Regional Remote Sensing Service Centre of space, Government of India located in the CAZRI campus for all the districts for preparation of Geographical Information System for systematic study and analysis. A critical review has been done on the drought proofing measures and their usefulness for the people and the area as well as to assess the severity of the drought in the areas. Vegetation is the only land category known to strongly absorb visible light and spectral reflectance of other material like soil and litter increase with wave length. The vegetation index lag correlation with rainfall use efficiency in time and space making direct satellite monitoring has been attempted for reliable and objective monitoring of drought conditions.

Critical analysis has been made for adjudging the severity of the droughts during past decades and its impact on the people and the environment. Case studies from different agro-climatic zones have been undertaken to assess the impact of drought in different regions. A questionnaire was prepared for detailed village level study to assess the impact of drought in different agro-climatic zones to know the intensity and severity of drought. The methodology for selection of districts, blocks and villages from each agro-climatic zones has been taken up in view of past tract of droughts in the different districts. The villages will be selected which were greatly affected by drought conditions in terms of time and impact on the area and the people. The research scholar had under undertaken the survey of about twenty villages to cover all the agro-climatic zones of the state. The information collected from villages was complied and analyzed to assess the prediction of droughts and immediate measures suggested to mitigate the impact to the possible extent.

2

ENVIRONMENTAL SETTING

Rajasthan is the largest state of India spread over in 3,42,239 sq kms of area and situated in the north-western part of the Indian Union between 23°30' and 30°11' North latitude and 69°29' and 78°17' East longitude. The state is largely an arid state for most of its part. The Tropics of Cancer passes through south of Banswara town. In physical formation it presents an irregular rhomboid shape, having maximum length of 869 kms from west to east and 826 kms from north to south. The western boundary of the state is part of the Indo-Pak international border running to an extent at 1070 km, touching four main districts of the region, viz. Barmer, Jaisalmer, Bikaner and Ganganagar. The state is girdled by Punjab and Haryana states in the north, Uttar Predesh in the east, Madhya Pradesh in the south east and Gujarat in the south-west (Fig. 1 and 2).

2.1 PHYSICAL CHARACTERISTICS

Physiography of Rajasthan is the product of long years of erosional and depositional processes. The present landforms and drainage systems have been greatly influenced and determined by the geological formations and structures. Four major physiographic regions can be delineated within the state viz. the western desert the Thar, the Aravalli hill region, the eastern plains and the south-eastern plateau region. The Aravalli hill ranges running from north-east to south-west, divides the state approximately into the western arid and eastern semi-arid regions and also a major water divide. The area in the east in well drained by integrated seasonal drainage systems, while the area to the west has only one integrated drainage system- the Luni system in the south-eastern part of the desert (Fig. 3).

2.1.1 The Western Desert

The western region is characterized by hyper-arid and arid landscape- barren hills, level rocky structural plains, rock-cut pediments, gravelly pavements, shallow colluvial plains, sandy plains with alluvium underneath, sandy hummocks, low sand

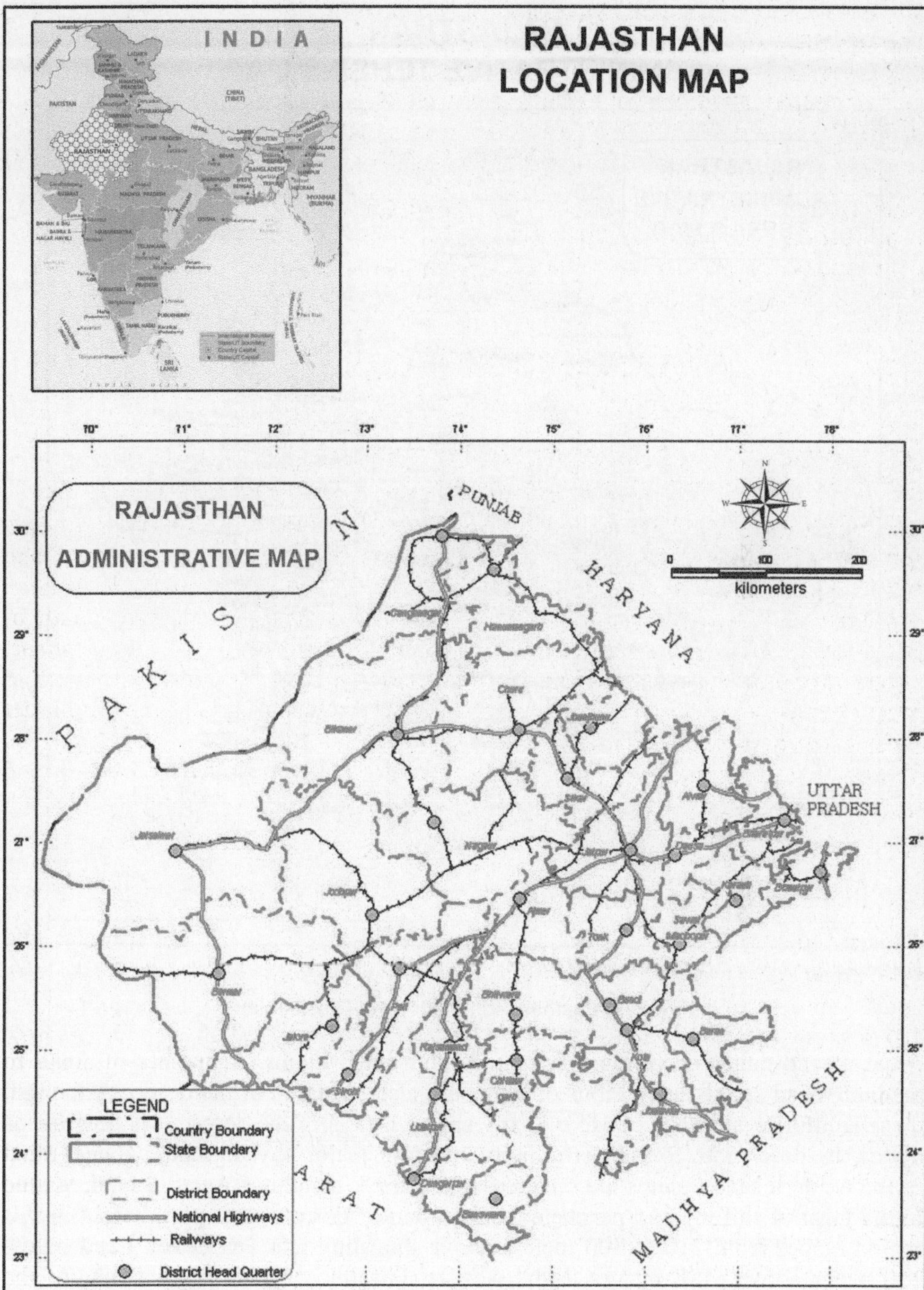

Source: Survey of India (SOI)

Fig. 1: Location Map

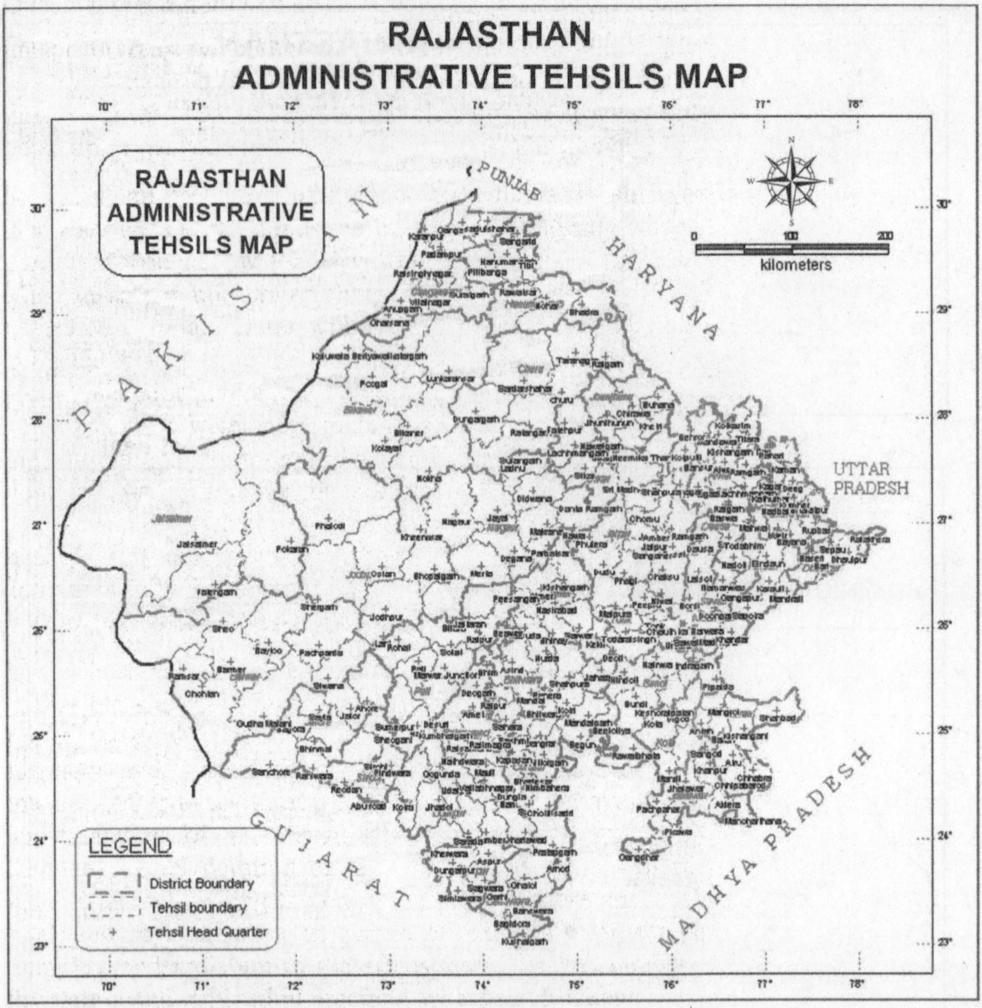

Source: Survey of India (SOI) and Census of India

Fig. 2: Rajasthan Administrative Tehsils Map

streaks, sand dunes of various kinds and inter-dunal plains. A number of small to medium sized saline depressions or ranns are also scattered in the desert. Although, these landforms are not exclusive to any specific parts of the desert, it is possible to divide the desert into four specific parts into four major physiographic zones. Most of the western sandy plains are covered with a thick mantle of Aeolian sand, visible in the form of shifting and permanent sandy dunes. For the most part, height in this region ranges from 210 to 300 meters above the Minimum Sea Level. Land below 150 meter above MSL occurs to the west of Jaisalmer. The western flanks of the Aravalli ranges are covered with sand deposits and the altitude varies from 300 to 600 meter above MSL. Four major physiographic zones formed in this region are narrated hereunder (Fig. 4A, B and 5A, B, C).

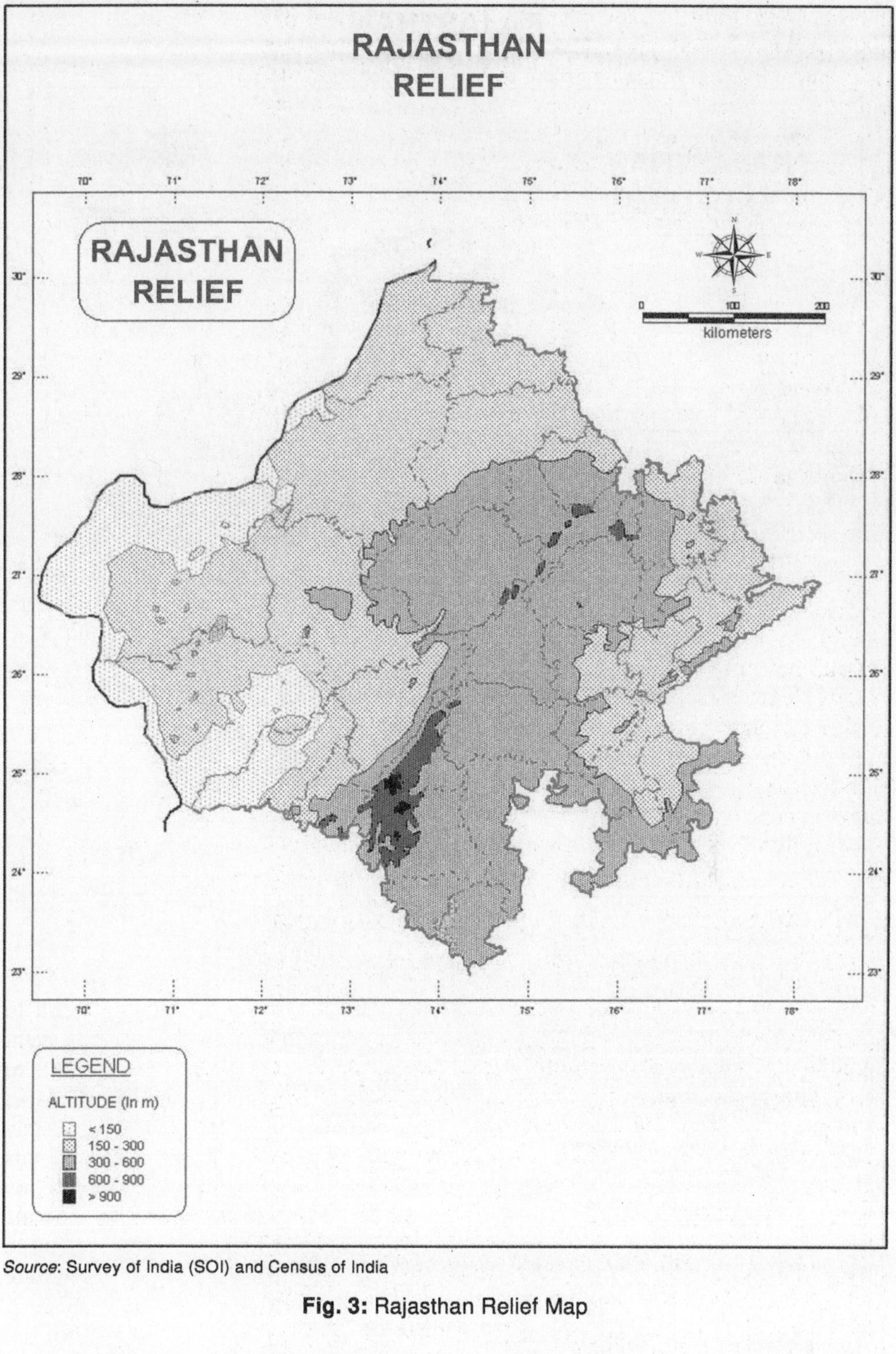

Source: Survey of India (SOI) and Census of India

Fig. 3: Rajasthan Relief Map

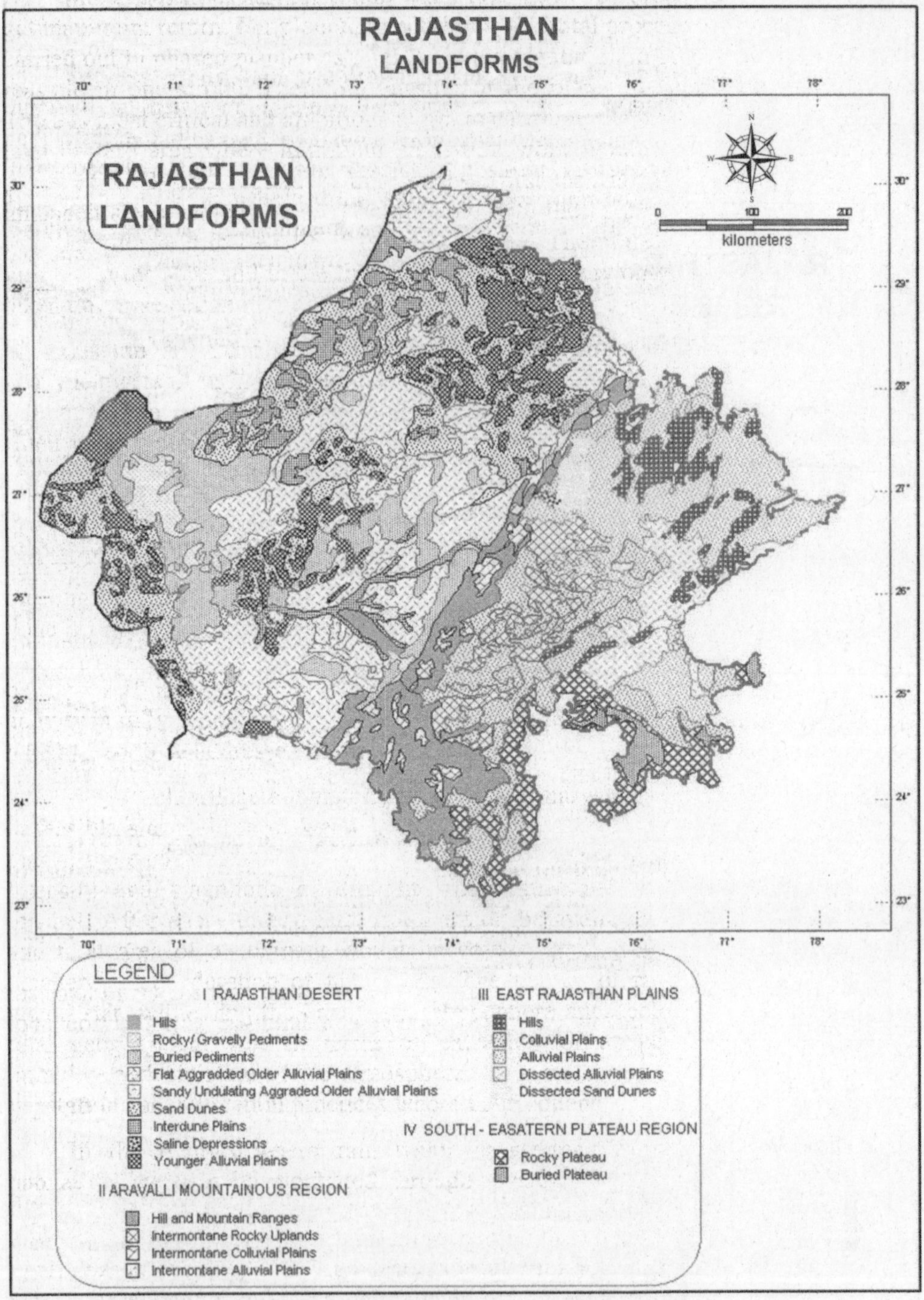

Source: Survey of India (SOI), NRSA, Mines and Geology Department, Udaipur and CAZRI, Jodhpur

Fig. 4A: Rajasthan Landforms Map

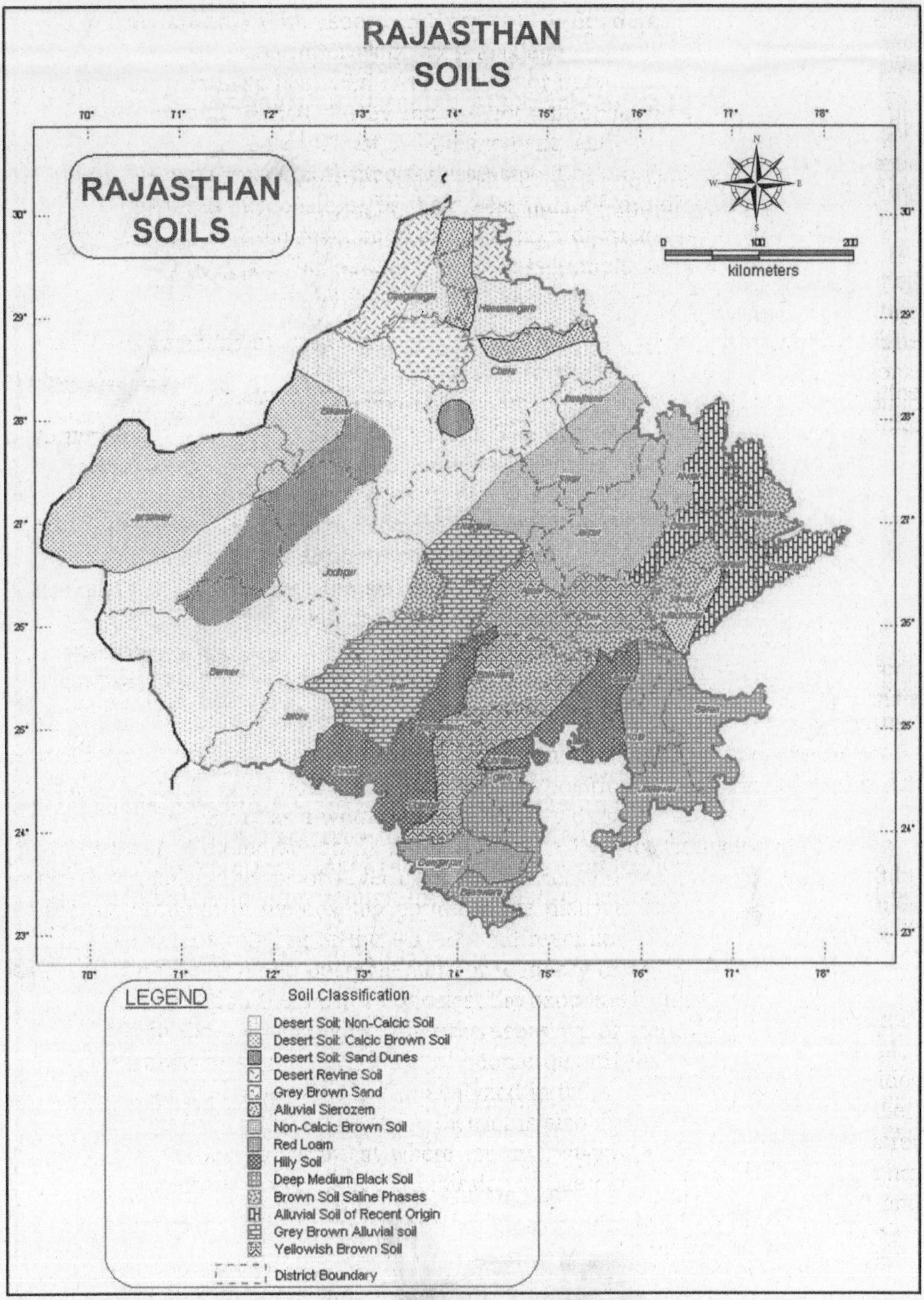

Source: Survey of India (SOI) and Department of Agriculture, Rajasthan

Fig. 4B: Rajasthan Soils Map

Source: Survey of India (SOI) and Department of Irrigation, Jaipur

Fig. 5A: Rajasthan Physiography and Drainage

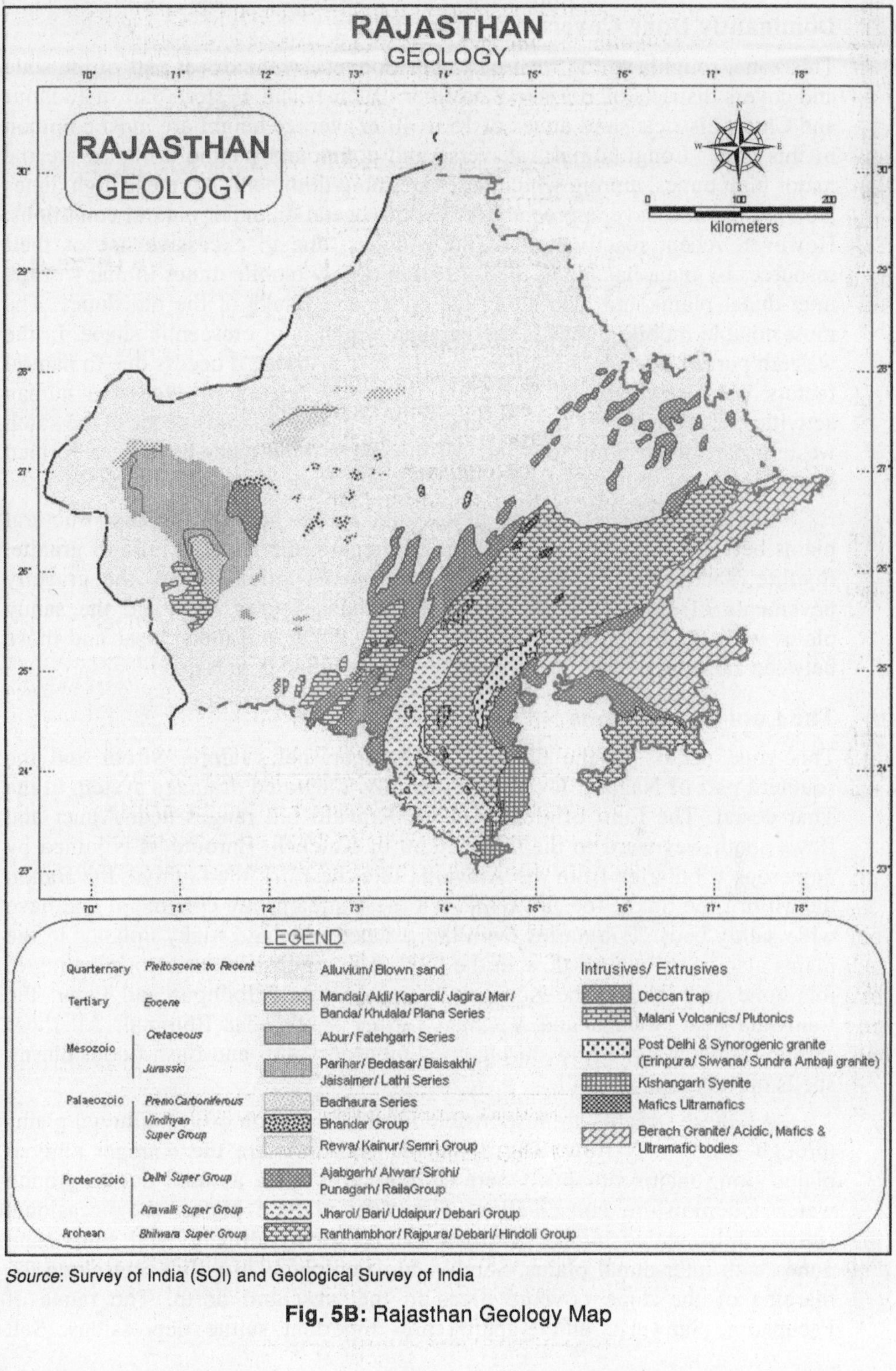

Source: Survey of India (SOI) and Geological Survey of India

Fig. 5B: Rajasthan Geology Map

(a) Dominantly Dune Covered Western Thar

This zone, roughly to the west of 25 cm isohyets, is the driest part of the state and covers districts of Barmer, Jaisalmer, Bikaner and western part of Jodhpur and Churu districts sand dunes of 15 to 40 m average height are most common in this zone. Longitudinal transverse and compound parabolic dunes are the major high dunes, among which latter are more dominant. All these high dunes are very old and have appreciable cover of vegetation under natural conditions. However, recent reactivation of these dunes, due to excessive use of their resources by man, has led to the formation of low mobile dunes in many of the inter-dunal plains and also along the crests and flanks of the old dunes. The most notable mobile dunes is the barchan which is of crescentic shape. In the western part of Barmer, Jaisalmer and Bikaner districts, it occurs due to natural factors but elsewhere, its formation is mostly related to excessive human activities. Chains of very high barchans or mega-barchanoids occur in the south western part of Jaisalmer district. All the above dune types have been formed by the dominant south-west wind, which blows from March to September.

Other dominant landforms in this zone are the high level rocky structural plains between Pokaran, Jaisalmer and Ramgarh, the isolated hills of granite, rhyolite, sandstone, limestone, rocky pediments around them, the gravelly pavements of Bap- Phalodi area and Chandan-Saukra area and the sandy plains with shallow soils. The ranns of Bap, Pokaran, Lunkaransar and those between Jaisalmer and Ramgarh are the major saline depressions.

(b) The Luni-Jawai Plains

This zone occurs in the districts of Jodhpur, pali, Jalore, Sirohi and the southern part of Nagaur. It contains the only integrated drainage system in the Thar desert. The Luni originates in the Aravalli hill ranges near Ajmer and flows south-westward to the Great Rann of Kutchch. Enroute, it is joined by numerous tributaries from the Aravallis like the Lilri, the Guhiya, the Bandi, the Mithri, the Sukri, the Jawai etc. All these streams are ephemeral and have wide sandy beds. Tributaries from the isolated hills and rocky uplands in the plains also contribute to flow in the Luni. The major tributaries originating in this zone and joining the Luni are two Johris near Jodhpur and Pipar, the Luniwala near Siwana and the Sagi further south near Bhinmal. All these streams register some flow during good monsoon rains and flash floods during spells of very high rainfall.

All these streams are responsible for the formation of vast alluvial plains through which they flow. The dominant landforms are the younger alluvial plains along major streams, where cultivation is more assured due to ground water and moisture conservation. The older alluvial plains, the occasional isolated hills with fringing pediments and buried pediments and some sand dunes with inter-dunal plains. Sandy undulated plains usually occur along the margins of the dune covered areas in the west and north. The ranns of Pachpadra, Sanwaria and Kaparda are important saline depressions. Salt affected alluvial plains are more common near the confluence of the Luni with the Great Rann.

(c) Sekhawati Tract of Low Dunes and Sandy Plains

The Shekhawati tract consists of Jhunjhunu, Sikar, Churu and northern part of Nagaur district. This zone is infested by sand dunes of low to medium height over the most of the area as well as deep sandy plains. Unlike of Luni-Jawai plains, this zone has very few streams most of which disappear after flowing for short distances of sandy plains. Import streams of this tract are Mendha and Rupangarh which originate in the Aravalli hill ranges and drain into the Sambhar salt lake. Another major stream, the kantli flows north-westward from the Aravallis and ultimately disappear in the sandy plains near Rajgarh. Sand dunes of this tract support more vegetation than the west due to higher rainfall. Along eastern fringe, many dunes are also noticed. Dissected by rills and gullies. Parabolic dunes cover maximum area but longitudinal and transverse dunes are also noticed. Dissected obstacle dunes occur along the foothills of Aravalli ranges and at the foot of the other isolated hills. The sandy plains usually have high fence line sandy hummocks along the boundaries of the cultivated fields. Many of the inter-dunal plains have developed saline depressions or ranns. Most notable are near Didwana, Sujangarh, Tal Chhapar, Parihara and Kuchaman. The Sambhar salt lake is situated astride the structural trend of the Aravallis and marks the eastern margin of the desert. Many of such ranns are used for production of salt on commercial basis.

(d) The Ghaggar Flood Plains

This zone of vast alluvial plain occurs in the north of the desert and occupies much of the area in Ganganagar and Hanumangarh districts. The wide and dry valley of Ghaggar is known to be the legendary Saraswati river which originated from Himalayas. The dry valley passes through Hanumangarh, Suratgarh and Anupgarh before entering into Pakistan in the west. Dry valleys of the Drishadvati, a major tributary of Saraswati, occur in Nohar-Bhadra area in the eastern part of Ganganagar district. Canals have been constructed through all the dry valleys making land around them very productive. The important canal systems of this tract are Gang canal system, Bhakra canal system, west Yamuna canal system and Indira Gandhi canal system. Now this area is seriously infested with water-logging and salinity problems at many places, specially between Suratgarh and Anupgarh. Low sand dunes and sandy hummocks are scatteredly distributed on deep sandy alluvial plains. Barchans have appeared in some parts of the plain region due to deep ploughing of soil without assured supply of canal water.

2.1.2 Aravalli Hills

Aravalli hill ranges constitute most dominant hilly area of Rajasthan. The ranges run diagonally across the state from north-east Delhi to south-west up to the plains of Gujarat covering a distance of about 690 kms. Within Rajasthan, the range runs from Khetri in the north-east to Khed Brahma in the south-west for a length of about 550 kms. In the north-east, the hill ranges become more prominent near Khetri and Alwar. Towards the south-west, the ranges become more prominent with peaks up to 1055 meter above MSL. Quartzite being very resistant, form most of the hills in the northern and central parts. Granite forms the high hills in the northern and central parts. Granite forms the high hills in the south near Abu. Apart from the hills, other major landforms within this region are rocky uplands, shallow to moderately deep colluvial plains and narrow alluvial plains at few locations.

The Aravalli ranges are the most prominent hill features extending from Sirohi, Udaipur and Dungarpur districts on the south-west to Jaipur and Alwar districts in the north-east. They rise to their highest summit of 1722 m above MSL at Mount Abu in Sirohi district. These ranges form a Labyrinth of low hills in Udaipur, Dungarpur and Banswara districts and stretch north eastwards in the form of undulating low hills through parts of Ajmer, Tonk, Sawai Madhopur, Karauli, Jaipur and Alwar districts. Isolated outcrops of Aravallis occur in Jhunjhunu, Sikar and Nagaur districts.

2.1.3 The Eastern Plains

Covering most parts of Alwar, Bharatpur, Jaipur, Dholpur, Tonk, Sawai Madhopur, Karauli, Bundi Baran, Jhalawar and Kota districts, the eastern plains have rich alluvial soils drained by seasonal rivers. Kota, Baran and Jhalawar districts have rocky surface structure with very shallow cover of black cotton soil, which is protected from soil erosion. The eastern plains are divided into Banas plains and Chappan plains in view of varied physical formations.

(a) Banas Plains

The region is a proad plain having an altitude of 150 to 300 m above MSL with slope towards east. This plain is drained by the Banas river through the northern part of Chittorgarh, Rajsamand and Udaipur districts and by several of its tributaries. Banas river catchments cover Bhilwara, Tonk, Ajmer, Jaipur, Dausa and Sawai Madhopur. Tributaries of Banas are Kothari, Dain, Sohadra on the left bank side while Berach, Wagon and Ghambir join in on the right bank side. Banas is major tributary of river Chambal which ultimately joins river Yamuna. Alluvium deposits become thin towards west where the plains are higher and more irregular, while the in the east, thickness of alluvium increases significantly.

(b) Chappan Plains

This plain lies in the south-eastern part of Udaipur, Rajsamand, Banswara, Dungarpur and southern part of Chittorgarh district. The area is trained by tributaries of Mahi river which ultimately reaches the Arabian sea through the Gulf of Cambay. The land in the south is deeply dissected resulting in the formation of separate hillocks.

2.1.4 South-Eastern Plateau

Southern and south-eastern Rajasthan s mostly a plateau called Hadoti plateau, having intrusions of black volcanic rocks into the Vindhyans, extends to a great part in Jhalawar, Baran and Kota districts. The Malwa plateau also extends in to the southern parts of Chittorgarh and Banswara districts. Average altitude of this plateau is 500 m above MSL and is dotted with isolated low ranges at a few places. This plateau in Rajasthan occurs in the upper catchment of the Chambal river in the south-east of the Mewar plains. The greater part of this area is drained by the river Chambal and its right bank tributaries like Kali Sindh, Parwan and Parvati. Physically, this plateau has been divided into two defined units- Vindhyan scrap land and Deccan lava plateau.

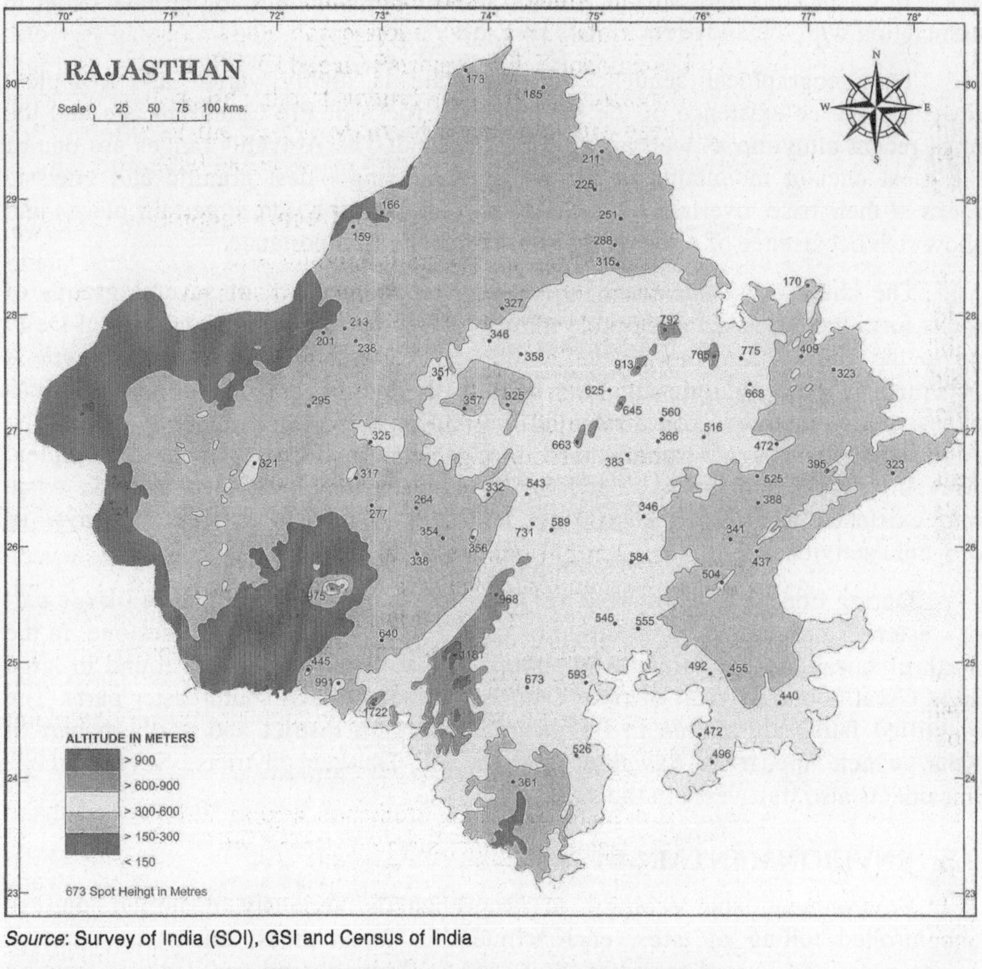

Source: Survey of India (SOI), GSI and Census of India

Fig. 5C: Rajasthan South-Eastern Plateau

2.2 GEOLOGICAL FACTORS

Rajasthan is endowed with a continuous geological sequence of rocks from the oldest Archaean metamorphites represented by Bhilwara Super Group from more than 2500 million years old to sub-recent alluvium and wind blown sand. The western and north-western parts of the state are covered by vast blanket of young unconsolidated deposits including the blown sand of the Thar Desert of western Rajasthan. The remaining area exposes wide variety of hard rocks which include various types of metamorphic schists, quartzites, marbles and gneisses of Pre-Cambrian age with associated acid and basic intrusive rocks. The sedimentaries includes rocks of Aravalli Super Group, Delhi Super Group, Upper Precambrian Vindhyan Super group, Cambrian to Jurassic, Cretaceous and Territary ages. The south-eastern extremity of the state is occupied by a pile of Basaltic flows of Deccan Traps of Cretaceous age. Several mineral deposits of economic importance occur in association with the above rock units (Fig. 5B).

The geographical sequence of the state is highly varied and complex, revealing the co-existence of the most ancient rocks of Pre-Cambrian age and the most recent alluvium as well as wind blown sand. The Aravallis ranges are one of the most ancient mountains in the world possessing oldest granitic and gneissic rocks at their base, overlain by rocks, highly metamorphosed at certain places and show rich occurrence of minerals of vital commercial importance.

The characteristic features of the geology is presence of several groups of rocks form the Aravalli mountain system which across the state from north of Delhi up to the Gulf of Cambay in the south-west. Central past of Aravalli ranges is occupied by great synlinorium. Due to thin deposits of sand in this region, rocks exposed are good but often engulfed in sandy alluvium and desert sand. The Archaeans consists of Bundelkhand Gneiss of the Banded Gneissic complex. Aravalli ranges are abnormally enormously thick series of argillaceous rocks came into existence when sediments were deposited in seas, underwent upheaval of orogenic activities.

During Eocene times marine transgression seems to have inundated large part of western Rajasthan with deposition of thick beds of fossilliferrus limestone. In the north of Jaisalmer, the Jurassics overlapped by various reasons. are found in large area/ Great boundary fault of river Chambal passes through south-easter parts. The identified faults are visible in Begun of Chittorgarh district and northern part of Kota, which appear in Sawai Madhopur and Dholpur districts. Several mega lineaments also transverse in the state.

2.3 ENVIRONMENTAL SETTING

Environment and ecology of Rajasthan is seriously disturbed due to uncontrolled felling of trees, encroachment in forest areas, mismanagement of scarce water and loss of moisture. Soil erosion through wind and water is creating land degradation conditions. Rainfall is erratic and uncertain, causing serious problems of cultivation and drinking water. Dryness in environment has been increased to great extent and causing difficult to vegetation. Uncontrolled grazing puts adverse impact on monsoon vegetation and most of the land surface has become denuded for clearance of land for cultivation requirements. Ground water depletion and excessive use of chemical fertilizers and pesticides have further intensified the environmental problem. Variation of rainfall is significant from normal annual precipitation as evident from Table 2.1.

Table 2.1: Annual Rainfall in Rajasthan

(In Centimetres)

District	Normal Annual Rainfall	Occurrence of Rainfall during the year				
		1999	2000	2001	2002	2003
1. Ajmer	60.18	35.74	37.95	56.36	19.09	47.65
2. Alwar	65.73	51.31	45.73	56.91	23.70	84.68
3. Banswara	95.03	66.13	49.10	64.44	52.46	84.03
4. Baran	87.38	92.91	80.19	89.89	32.21	71.47
5. Barmer	26.57	21.00	26.36	29.81	8.75	47.83
6. Bharatpur	66.39	62.34	38.30	59.62	39.46	75.19
7. Bhilwara	68.32	50.00	46.64	70.41	31.46	58.49
8. Bikaner	24.30	26.38	31.13	29.40	6.63	29.95
9. Bundi	77.34	66.18	57.61	80.92	35.29	69.25
10. Chittorgarh	84.15	76.20	47.76	70.34	41.00	67.34
11. Churu	35.47	22.39	24.00	37.18	16.22	41.37
12. Dausa	56.10	53.64	44.64	58.42	30.49	79.12
13. Dholpur	74.45	62.00	48.11	66.51	46.59	62.78
14. Dungarpur	72.89	53.03	40.10	59.74	38.06	67.20
15. Ganganagar	22.64	16.02	12.97	29.14	8.90	25.94
16. Hanumangarh	27.35	22.70	16.76	37.96	13.56	29.83
17. Jaipur	56.38	43.15	37.59	43.09	20.74	64.06
18. Jaisalmer	18.55	27.27	18.53	21.96	4.40	21.78
19. Jalore	37.00	26.17	31.18	36.20	17.30	72.31
20. Jhalawar	84.43	114.21	72.61	96.86	48.44	74.90
21. Jhunjhunu	40.51	32.08	34.00	46.48	16.93	52.51
22. Jodhpur	31.37	31.32	23.67	37.53	8.77	36.99
23. Karauli	67.07	58.01	46.25	54.43	26.47	78.08
24. Kota	73.24	85.57	76.92	97.07	38.63	63.57
25. Nagaur	31.17	24.67	31.64	36.06	14.42	46.91
26. Pali	42.44	45.27	33.74	55.49	17.60	54.98

District	Normal Annual Rainfall	Occurrence of Rainfall during the year				
		1999	2000	2001	2002	2003
27. Rajsamand	56.78	45.41	30.73	62.31	30.09	49.93
28. Sawai Madhopur	87.34	64.34	38.98	64.54	23.95	66.11
29. Sikar	44.03	27.89	27.87	45.89	19.80	47.70
30. Sirohi	59.12	39.43	51.24	74.26	27.04	92.27
31. Tonk	66.83	64.37	50.11	57.72	23.97	62.64
32. Udaipur	64.50	43.42	39.99	63.33	38.77	65.80
Average- Rajasthan	**57.51**	**50.56**	**41.64**	**57.43**	**26.71**	**60.47**

Source: Statistical Abstract, 2003, Rajasthan

Analysis of rainfall during last five years reveals that there is significant variation in the occurrence in different parts of the state. Year 2002 remained worst year and 2001 and 2003 were good years, while 1999 and 2000 were deficient years. Inter-district variations are vital in view of regional problems as western Rajasthan districts are getting less than 50 percent than the average rainfall. Condition of Jaisalmer district is worst with average rainfall of 18.55 cms and 4.40 cms during 2002. Normal rainfall or total annual rainfall is not the measuring indicator for cultivation practices in view of timely rains with regular intervals of occurrence. Eastern and southern districts receive good rains but crop failure is recurrent feature due to uneven spells. Occurrence of rains in total duration of monsoon remains varied and erratic, which creates various environmental problems (Fig. 6 and 7).

Temperatures are other vital environmental features, which play significant role in vegetation. Highest temperature varies between 44 to 49 degree Celsius while minimum temperature ranges between (-)3 to 7 degree Celsius in various parts of the state. Humidity in percentage terms ranges between 36 to 67 in different regions and rate of evaporation is highest in summer months when the maximum temperature crosses the limit of 49 degree Celsius (Fig. 8 to 11).

2.4 SOCIO-ECONOMIC CHARACTERISTICS

Socio-economic conditions are governed by physical and environmental features. Rajasthan is basically agrarian economy where most of rural people are dependent on cultivation practices, which are largely dependent on monsoon. In addition to agriculture, animal husbandry is the second source of livelihood and in case of failure of rains, livestock helps the people to great extent in survival. Rajasthan being largest state in the country in geographical area, land resources are sufficient but deficiency of water is the main obstacle in socio-economic development of the state. About 50 percent people of the state are directly dependent on land for their livelihood, others are also indirectly related for wage employment, processing, marketing and with the purchasing power of cultivators for purchase of other commodities of their requirement. In case of monsoon failure, total economy of the state is seriously affected.

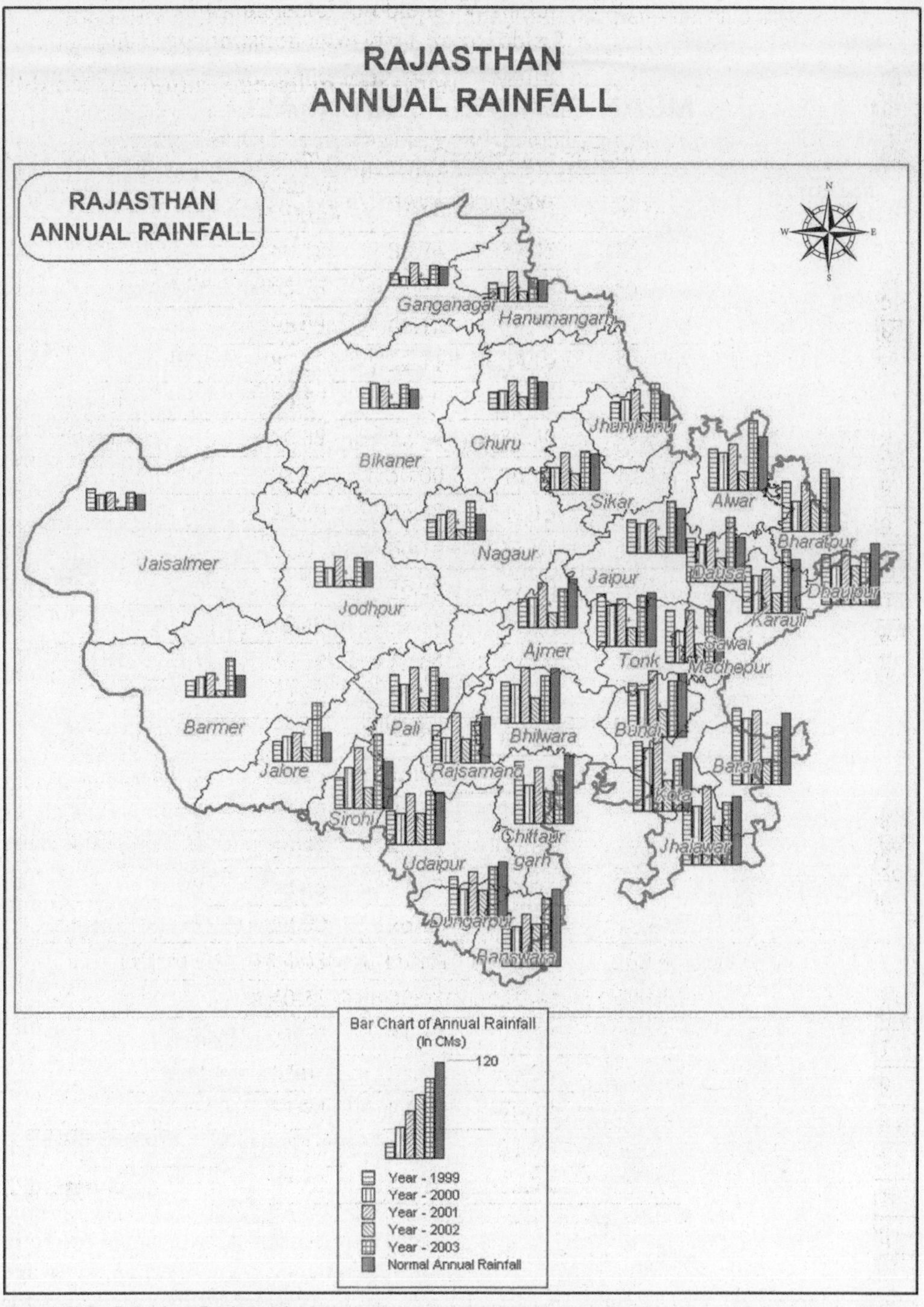

Source: IMD, Rajasthan Region, Jaipur

Fig. 6: Rajasthan Annual Rainfall Map

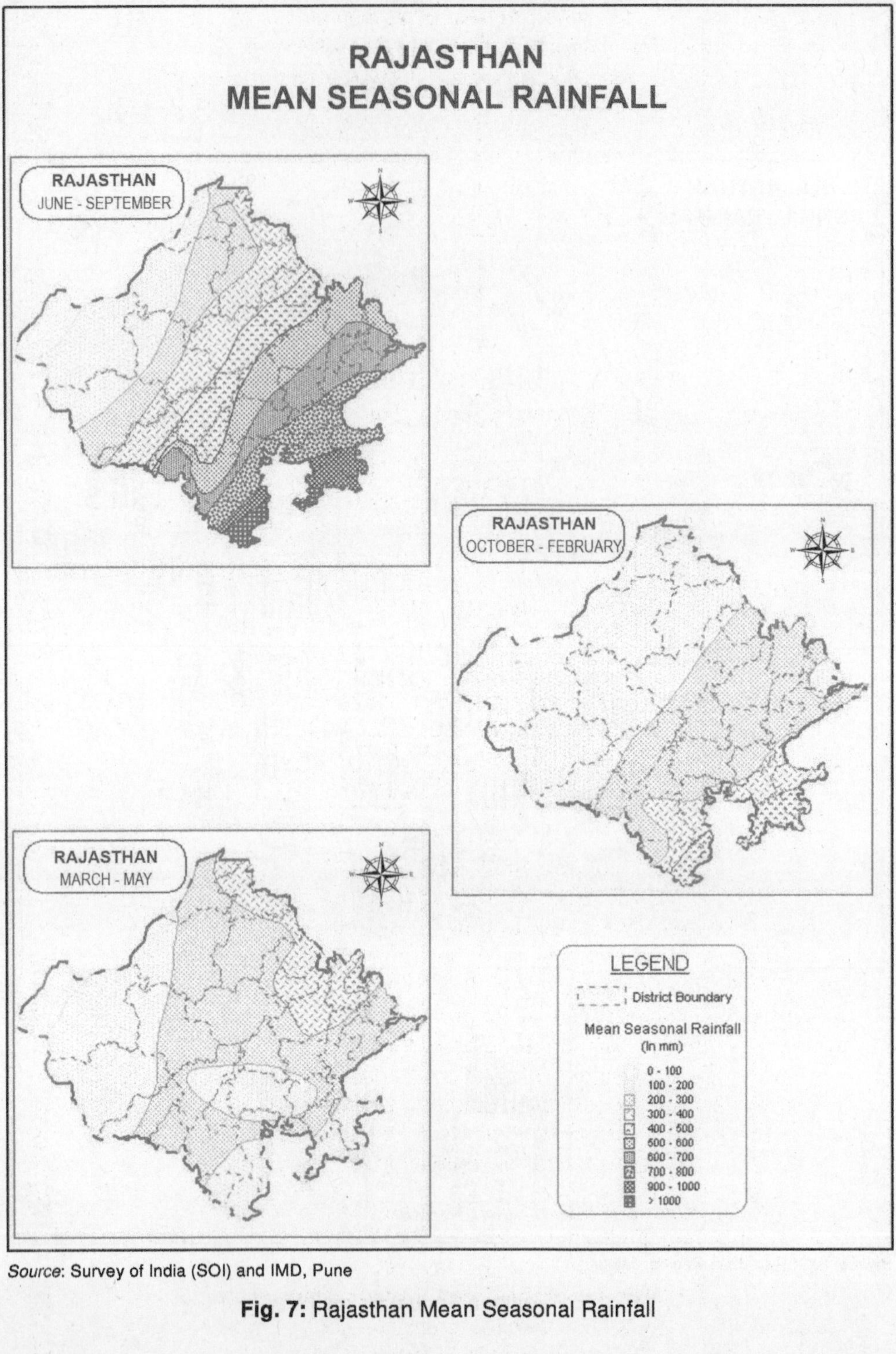

Source: Survey of India (SOI) and IMD, Pune

Fig. 7: Rajasthan Mean Seasonal Rainfall

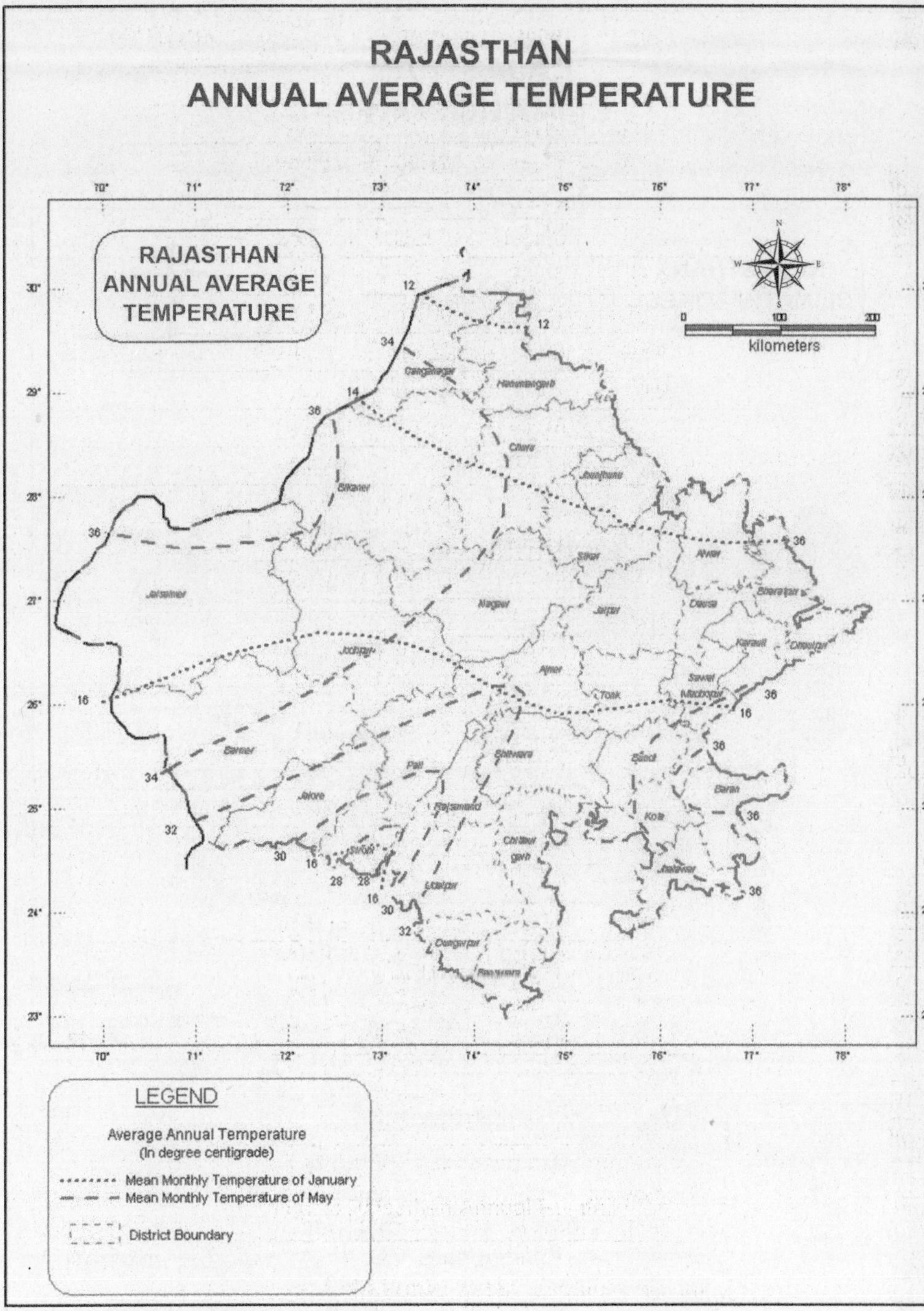

Source: Survey of India (SOI) and IMD, Pune

Fig. 8: Rajasthan Annual Average Temperature

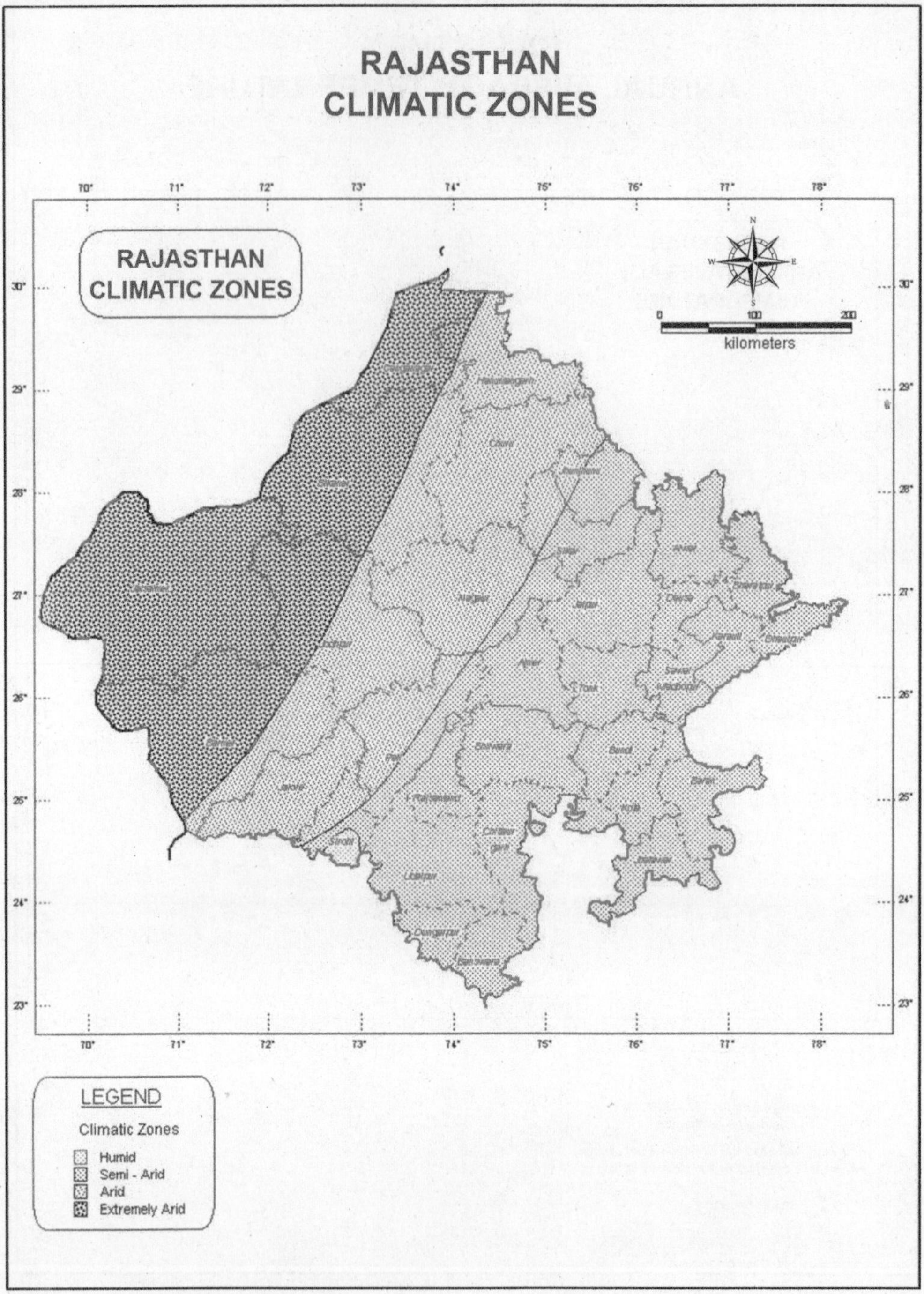

Source: Survey of India (SOI) and IMD, Pune

Fig. 9: Rajasthan Climatic Zones

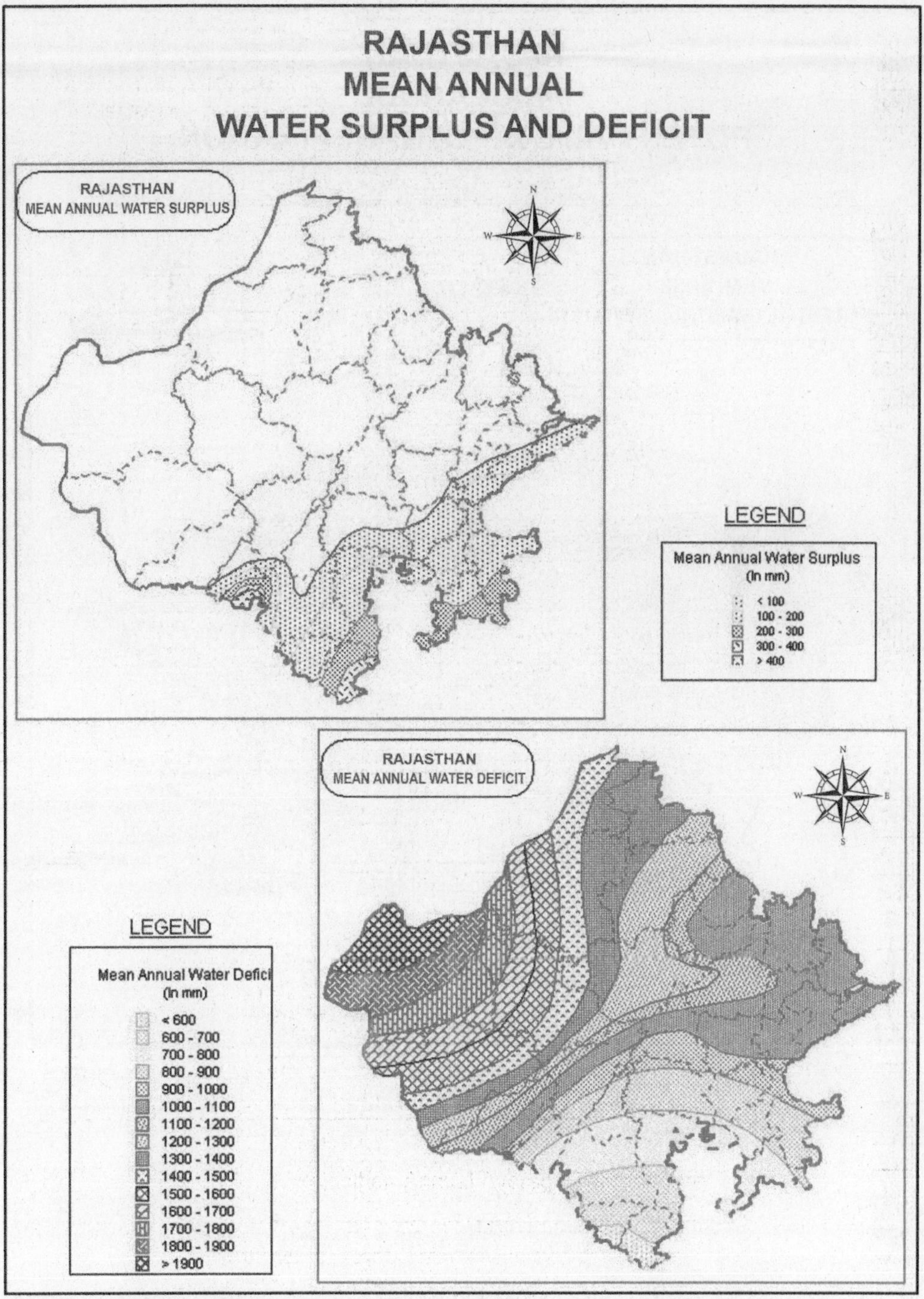

Source: Survey of India (SOI) and IMD, Pune

Fig. 10: Rajasthan Mean Annual Water Surplus and Deficit

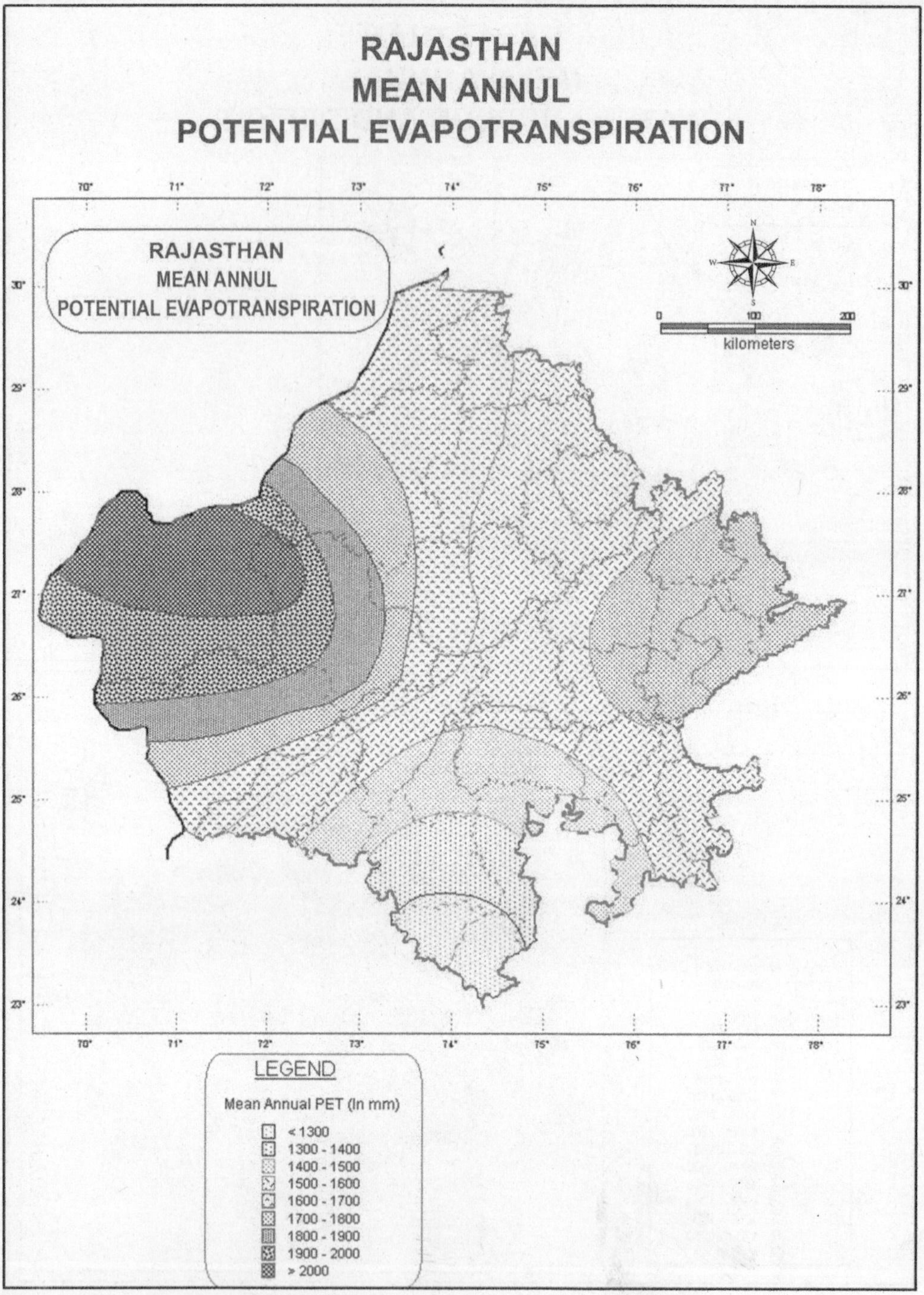

Source: Survey of India (SOI) and IMD, Pune

Fig. 11: Rajasthan Mean Annual Potential Evapotranspiration

2.4.1 Demographic Features

Rate of population growth is quite high in Rajasthan as population of the state was 103 lakhs in 1901 which increased to 565 lakhs. During the decade 1991-2001 rate of population growth of state was 28.41 percent which is significantly higher to country population growth of 21.52 percent in the same period. Inter-district variance in the same period are also varied between 20.0 to 47.5 percent. Even after tremendous population growth the state is trying to produce agriculture produce more than its requirement in normal conditions. Demographic features of state are given in Table 2.2.

Table 2.2: Demography Features of Rajasthan-2001

Districts	Area (Sq kms)	Population (Total)	Population (Rural)	Growth rate (%) 1991-01	Density Persons (per sq kms)	Sex Ratio
1. Ajmer	8481	2181670	1306994	26.2	257	931
2. Alwar	8380	2992592	2557653	30.3	357	886
3. Banswara	5037	1501589	1394226	29.9	298	974
4. Baran	6992	1021653	849638	26.1	146	909
5. Barmer	28387	1964835	1819431	36.9	69	892
6. Bharatpur	5066	2101142	1692182	27.2	414	854
7. Bhilwara	10455	2013789	1598938	26.4	192	962
8. Bikaner	27244	1674271	1079235	38.2	61	890
9. Bundi	5776	962620	783058	25.0	173	907
10. Chittorgarh	10856	1803524	1514255	21.5	166	964
11. Churu	16830	1923878	1387682	24.7	114	948
12. Dausa	3432	1317063	1181245	32.4	384	899
13. Dholpur	3033	983258	806640	31.2	324	827
14. Dungarpur	3770	1107643	1026787	26.7	294	1022
15. Ganganagar	10978	1789423	1336066	27.6	163	873
16. Hanumangarh	9656	1518005	1214467	24.4	157	894
17. Jaipur	11143	5251071	2659004	35.1	471	897
18. Jaisalmer	38401	508247	431853	47.5	13	821
19. Jalore	10640	1448940	1338946	26.8	136	964
20. Jhalawar	6219	1180323	1012081	23.3	190	926
21. Jhunjhunu	5928	1913689	1518573	20.9	323	946
22. Jodhpur	22850	2886505	1909423	34.0	126	907
23. Karauli	5524	1209665	1037720	30.4	218	855
24. Kota	5217	1568525	729948	28.5	288	896
25.Nagaur	17718	2775058	2297721	29.4	157	947

Districts	Area (Sq kms)	Population (Total)	Population (Rural)	Growth rate (%) 1991-01	Density Persons (per sq kms)	Sex Ratio
26. Pali	12387	1820251	1429364	27.5	147	981
27. Rajsamand	3860	987024	858301	20.0	256	1000
28. Sawai Madhopur	4498	1117057	904417	27.6	248	889
29. Sikar	7732	2287788	1815250	24.1	296	951
30. Sirohi	5136	851107	700217	30.1	166	943
31. Tonk	7194	1211671	958503	24.3	168	934
32. Udaipur	13419	2633312	2142995	27.4	196	971
Total- Rajasthan	**342239**	**56507188**	**43292813**	**28.4**	**165**	**921**

Source: Statistical Abstract, Rajasthan and Census of India

Rural and urban composition of population of the state is 76.61 and 23.39 percent respectively habitating in 41353 villages and 216 cities and towns. Density of population of the state is 165 persons per sq kms in comparison to country figure of 325 persons. Density of population is also varied as Jaisalmer district has density of population of only 13 persons per sq km while Jaipur the density of population is highest with 471 persons. Average rural population per village is 1047 persons and per city or town 61178 persons. Sex ratio of population is varied in the state 921 female per 1000 male in the state in comparison to country figure of 933. Lowest sex ratio of 821 female per 1000 male persists in Jaisalmer district and Dungarpur district has highest figure of 1022 female per 1000 male. This position also reflects in the social features of the state (Fig. 12, 13, 14 and 15).

Rajasthan state has limited water resources to cater all the needs of the state and even with low population density, pressure on land resources is quite high which is main feature of environmental degradation and loss of vegetation cover over the earth surface. The state is also deficient in other resources as creation of infrastructure facilities and basic amenities is costly in view of long distances specially in western Rajasthan where population is scattered in large area and in the districts of Jaisalmer and Barmer there is absence of village culture for want of water. People reside in close vicinity of water points which are unable to cater large sect of population. Eastern and southern part of the state is comparatively better than the western region in respect of water resources. However, total state is seriously facing the acute water shortage to accelerate socio-economic development.

People of the state are surviving in most severe conditions as there are various geographical and environmental adversities, creating serious hindrances in the development. High rate of population growth has further intensified the problems, as per capita cost of each development component is very high due to geographical conditions. People of the state are courageous enough to face all the adverse conditions and surviving with limited sources of livelihood.

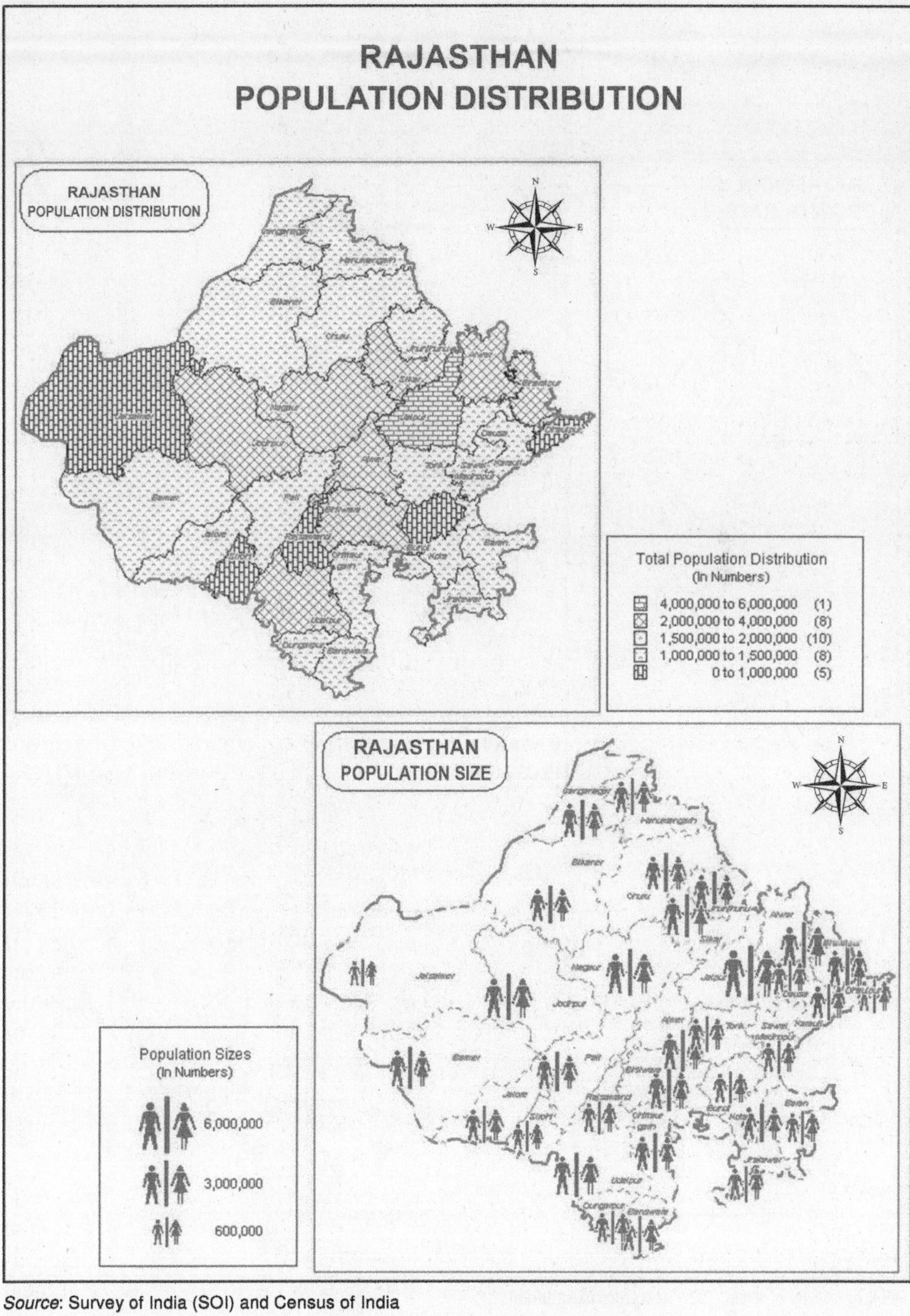

Source: Survey of India (SOI) and Census of India

Fig. 12: Rajasthan Population Distribution

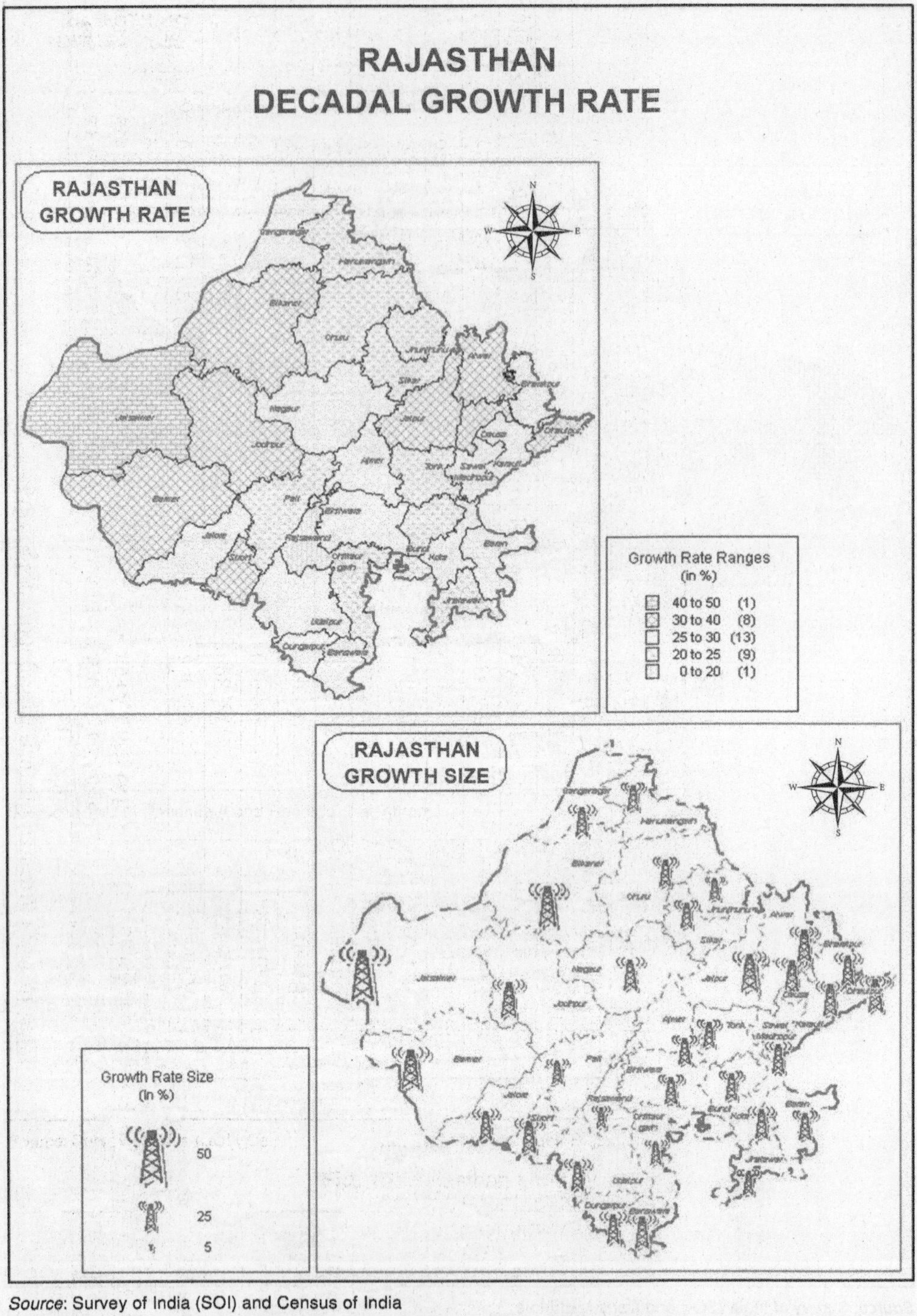

Source: Survey of India (SOI) and Census of India

Fig. 13: Rajasthan Decadal Growth Rate

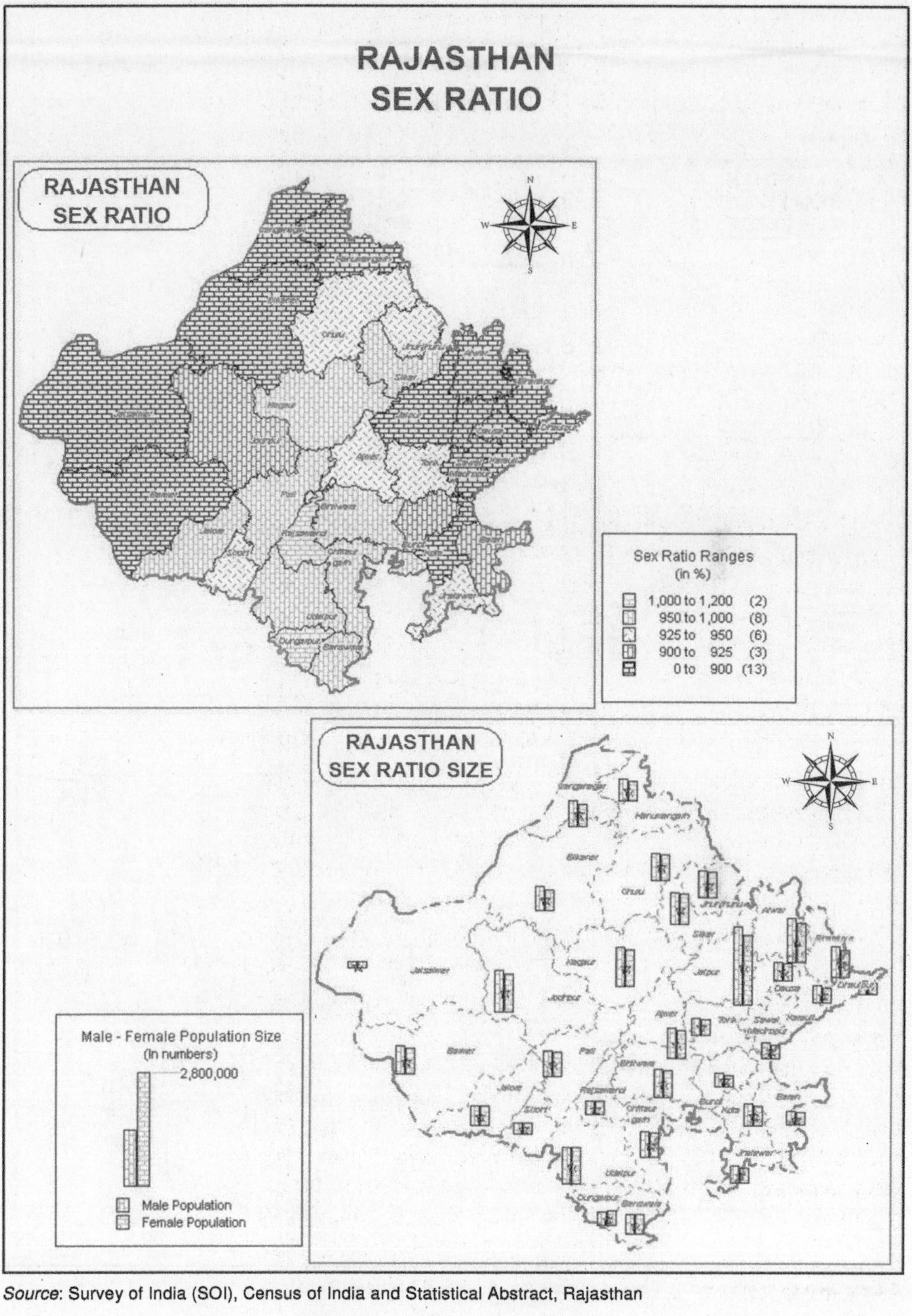

Source: Survey of India (SOI), Census of India and Statistical Abstract, Rajasthan

Fig. 14: Rajasthan Sex Ratio

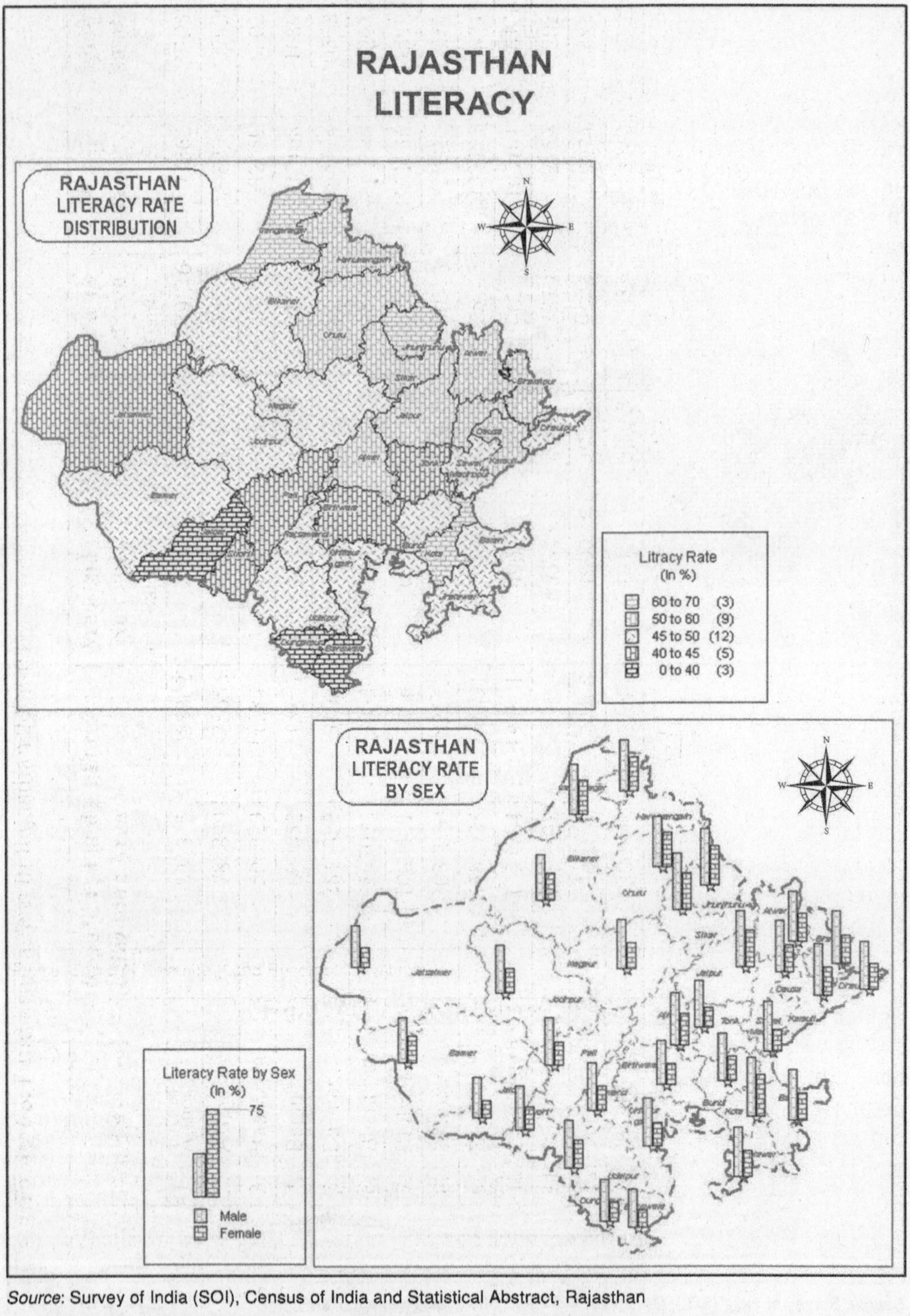

Source: Survey of India (SOI), Census of India and Statistical Abstract, Rajasthan

Fig. 15: Rajasthan Literacy

2.4.2 Land Use Pattern

Rajasthan state has vast geographical area and density of population is low specially in western region. Land use pattern of the state is evident that western Rajasthan, the Great Indian Thar desert is spread over in 61 percent of area covering 12 districts and 40 percent population whereas the eastern and southern part has only 39 percent geographical area spread over in 20 districts and having 60 percent of state population. Thus the population pressure on land resources in high in eastern and southern parts of the state but the western region is unable to bear even 40 percent of state population due to deficiency of natural and other resources.

Land use pattern is helpful to understand the utilization status of scarce land resources under various uses. This also helps to take suitable measures while assessing the nature of land remaining unutilized for various geographical and environmental features. Productivity of the land keeping in view the surface and sub-surface resources and deficiencies are helpful in taking short term and long term decisions to use the land resources in most appropriate manner. Land use pattern of Rajasthan under various categories is given in Table 2.3.

Analysis of land use pattern under different categories reveals that 7.73 percent area has been remained under forests, 5.15 percent of the land area is put under non-agricultural uses like habitations, industries, mines, roads and railways and other constructed areas, 7.34 percent area is barren land not fit for cultivation practices, 5.01 percent area is under permanent pastures, grazing land and tree groves, 14.20 percent area is wasteland of various categories. 29.03 percent area is lying as fallow land and 31.54 percent area is put under cultivation practices. The present land use pattern is result of various changes took place by continuously changing the land use pattern (Fig. 16A, B).

Area under forests undergone various changes due to regular encroachment by people for meeting cultivation requirements and uncontrolled felling of trees. Even in the present situation the Forests and Environment Department of Rajasthan claims 3248801 hectare area under its possession where 36.51 percent area is proclaimed as reserve forests, 54.33 percent area as protected forest area and 9.16 percent area under unclassified forests. There is significant difference between the figures of forests department and board of revenue and 18.40 percent area is not in possession of forests department. The encroachments are the regular feature and done with the connivances of powerful persons as encroachment in forest area is not permissible under Forests and Environment Act. In view of present circumstances, it can very well presumed that before 5000 years back the state was thick forests and human interference has squeezed the prestigious resources.

Table 2.3: Land use Pattern in Rajasthan 2002-03

(Area in hectares)

District	Reporting Land Utilization Area	Forests	Area under Non-agri. uses	Barren land	Pastures and tree groves	Waste land	Fallow Land	Net Area Cultivated
1. Ajmer	842345	55062	48874	89229	79716	78562	175386	315516
2. Alwar	782899	78656	45558	83134	24525	8296	70301	472429
3. Banswara	506279	112848	11155	60653	13024	32059	50279	226261
4. Baran	699652	215857	24046	41420	38382	22168	168974	188805
5. Barmer	2817332	31026	73680	127010	202918	266514	1055837	1060347
6. Bharatpur	507073	30334	29663	24114	8949	2986	22164	388863
7. Bhilwara	1047441	74063	66950	145748	120403	161036	159292	319949
8. Bikaner	3037030	82543	257761	35910	50533	823813	1526118	260352
9. Bundi	581938	141891	39676	50468	25443	33757	97679	193024
10. Chittorgarh	1035704	193841	49626	98384	76236	165101	61968	390548
11. Churu	1385898	6442	63343	677	38392	12154	720249	544641
12. Dausa	340467	24465	19909	17173	26987	9074	48400	194459
13. Dholput	300905	27059	16183	58463	18381	13042	27220	140557
14. Dungarpur	385593	61241	22971	73589	36854	22776	59252	108910
15. Ganganagar	1093352	59340	68253	2773	1960	38482	382188	540356
16. Hanumangarh	970315	15048	58625	705	4089	9252	207810	674786
17. Jaipur	1105494	81522	78265	50092	78764	34401	203845	578605
18. Jaisalmer	3839154	23291	118365	369064	103547	2594091	552904	77892
19. Jalore	1056602	20096	40331	83382	47404	25496	303124	536769
20. Jhalawar	632235	121206	25146	37618	52481	52214	60771	282799

District	Reporting Land Utilization Area	Forests	Area under Non-agri. uses	Barren land	Pastures and tree groves	Waste land	Fallow Land	Net Area Cultivated
21. Jhunjhunu	591683	39613	20205	15377	40351	6473	143874	325790
22. Jodhpur	2256405	6996	78920	145888	122641	74415	1697423	190122
23. Karauli	505217	171708	20197	52032	31226	14451	62869	152734
24. Kota	521133	121302	27797	38189	15004	28175	90601	200065
25. Nagaur	1764380	17989	85526	58047	73375	14821	777326	737296
26. Pali	1233079	80245	58139	145942	90860	49317	454636	353940
27. Rajsamand	455093	24663	23466	104601	57665	121603	36311	86784
28. S. Madhopur	499424	78605	34105	41986	25697	11946	150616	156469
29. Sikar	774244	60999	34282	17333	40917	10619	188054	422041
30. Sirohi	517947	154826	24626	75885	33415	9860	101149	118186
31. Tonk	717960	26915	42324	30752	44601	46921	185695	340752
32. Udaipur	1462105	411322	156615	338733	90955	132506	104916	227058
Rajasthan	**34266378**	**2651014**	**1764582**	**2514371**	**1715694**	**4866381**	**9947231**	**10807105**

Source: Board of Revenue (Land Records) Rajasthan

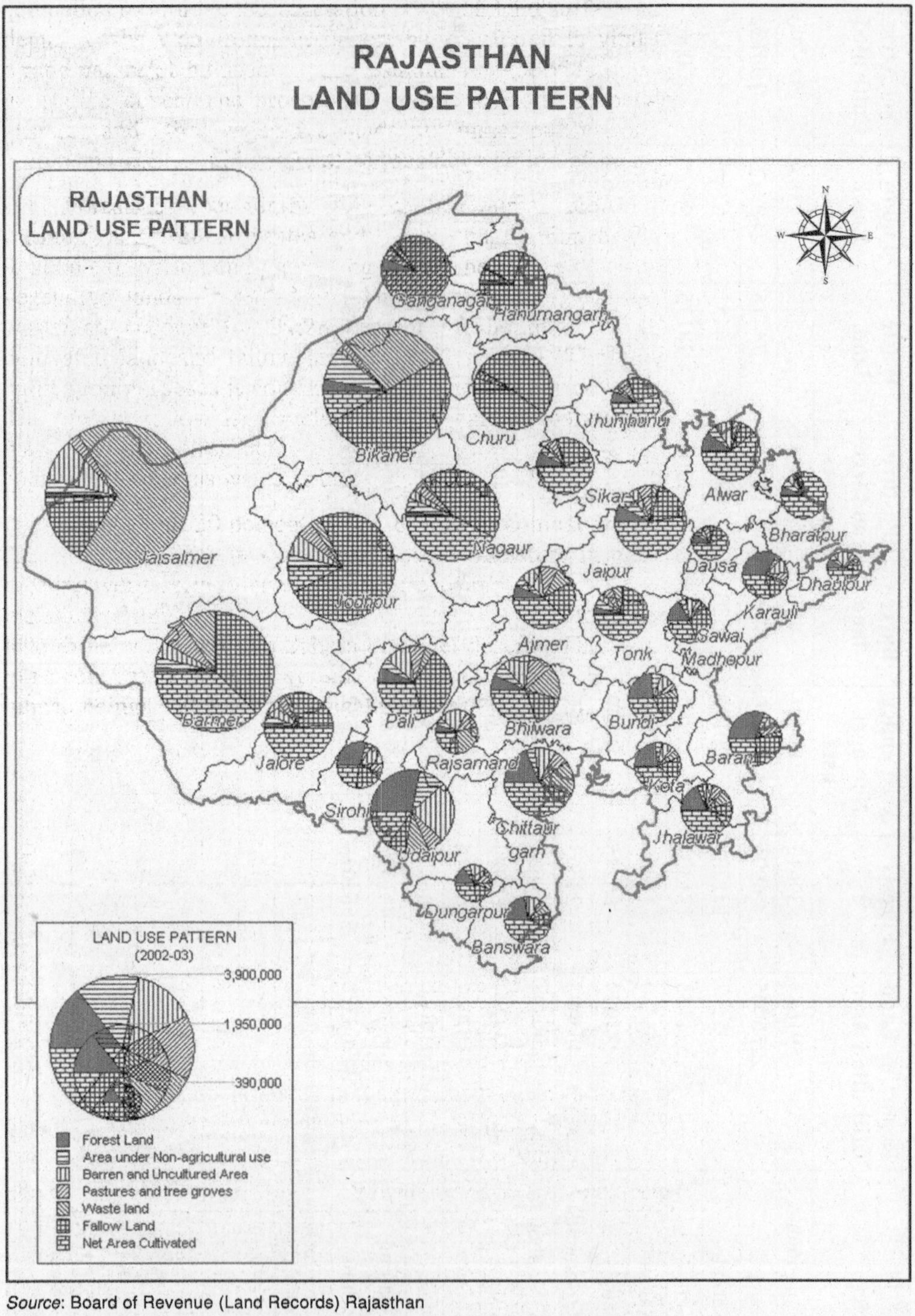

Source: Board of Revenue (Land Records) Rajasthan

Fig. 16A: Rajasthan Land use Pattern

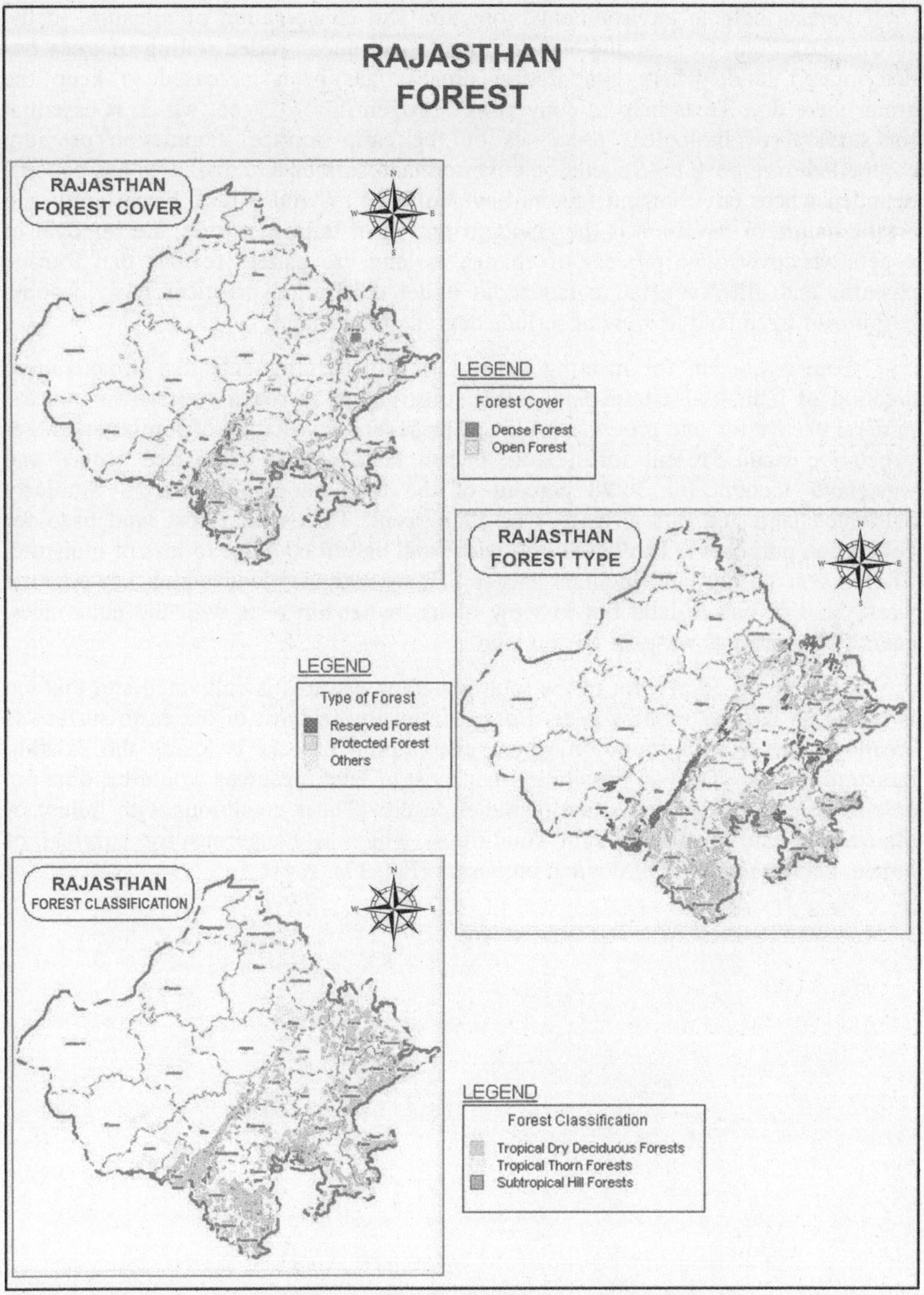

Source: Survey of India (SOI), Department of Forest, Government of Rajasthan and Forest Survey of India

Fig. 16B: Rajasthan Forest Map

Forests help in environment protection and conservation of moisture in the surrounding area. Continuous encroachments and uncontrolled felling of trees has deteriorated land, where evaporation process has been increased to keep the atmosphere dry. Trees help in converting nitrogen in to oxygen, which is essential for survival of biological processes on the earth surface. Population pressure compelled to remove the vegetation cover on land surface and total area has become denuded where environment factors have suffered to vital extent. Uncertainty and erratic nature of monsoon is the resultant impact of felling of trees and removal of vegetative cover. The process of change in land use pattern reveals that forests, pastures and other vegetative lands put under cultivation practices have become barren and wasteland in view of serious degradation of land.

Forests are cut for meeting various human requirements like urbanization, creation of industrial estates and areas, construction of roads and railway tracks, mineral excavation and processing units. The severe devastation of forests and other vegetative lands Present forest land, barren land, pastures and tree groves and wasteland account for 34.28 percent of the total geographical area. Similarly cultivated land and fallow land is 60.57 percent. Thus total forest land used for cultivation purposes is converted into waste and barren land due to loss of moisture. The present fallow land includes larger part of marginal land which has become waste land or barren land but in view of its ownership rests with the cultivators, cannot be termed as waste or barren land.

Thus about 30 percent of the total geographical area is cultivated and this too is losing its fertility in slow pace. Loss of moisture in most of the earth surface is creating very serious condition where evaporation process is losing the existing moisture of land. There is possibility that most of land resources would become not able to be cultivated like Sahara or Sahel deserts. These conditions with failure of monsoon create severe drought conditions, which is dangerous for survival of human beings and other biological processes (Fig. 17).

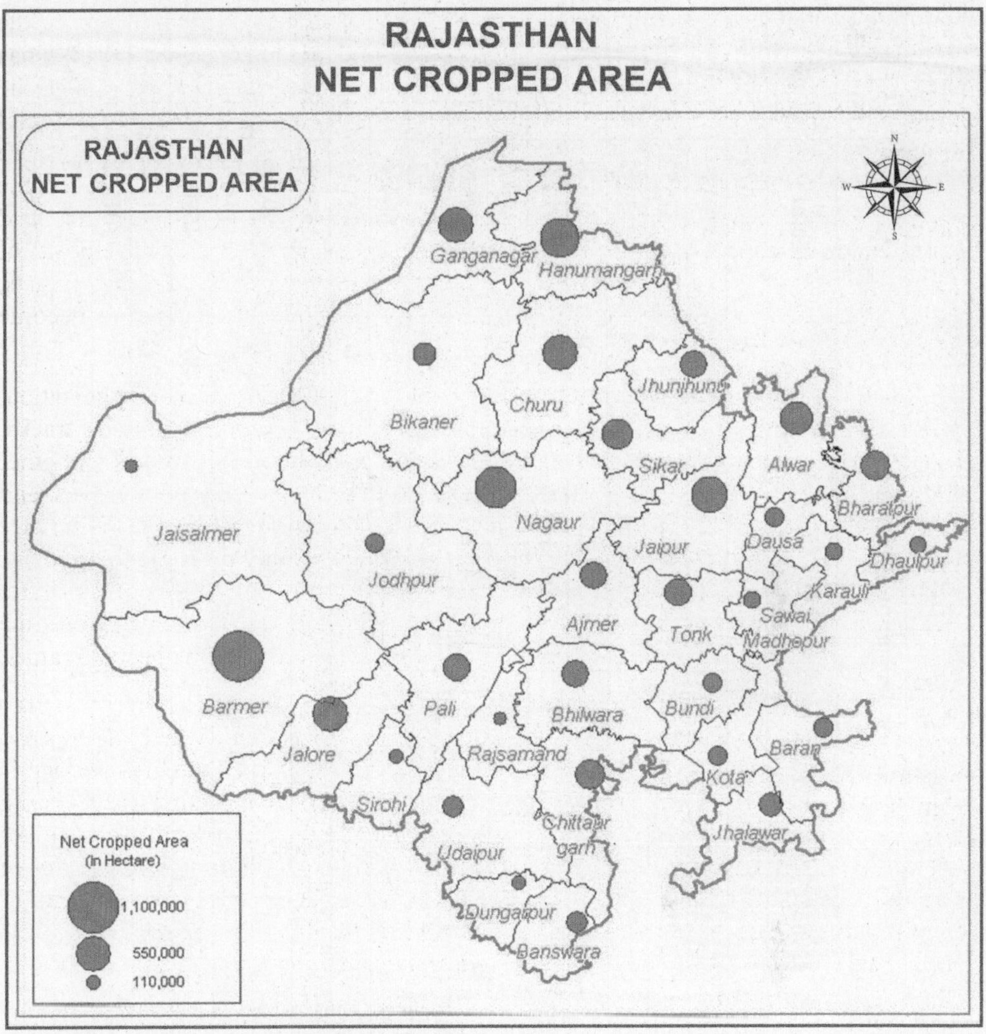

Source: Board of Revenue (Land Records), Rajasthan

Fig. 17: Rajasthan Net Cropped Area

2.4.3 Irrigation Facilities

Irrigation facilities help in use of land resources for two or more crops as one crop as usually done in various parts of the country. The position of Rajasthan state is altogether different as crop intensity is 122.30 percent irrespective of the fact that 40.45 percent of cultivated land is irrigated. The irrigation facilities continue to fluctuate due to availability of water from monsoon rains and surface irrigation systems. District-wise and source-wise irrigation facilities available during 2002-03 are given in Table 2.4.

Table 2.4: Area Irrigated by Sources

(In Hectare)

District	Net Cropped area	Irrigation by sources					Crop Intensity
		Canal	Tank	Wells & Tube Wells	Other Sources	Total	%
1. Ajmer	315516	5	7	33620	39	33671	105.34
2. Alwar	472429	—	—	410015	134	410149	141.55
3. Banswara	226261	38688	2403	9010	7451	57552	125.99
4. Baran	188805	2593	248	111104	4008	117953	127.30
5. Barmer	1060347	—	—	121604	20	121624	108.37
6. Bharatpur	388863	4626	—	292459	—	297085	139.20
7. Bhilwara	319949	3905	944	42705	1051	48605	113.85
8. Bikaner	260352	96667	—	43951	—	140618	119.74
9. Bundi	193024	—	205	101269	853	102327	135.57
10. Chittorgarh	390548	60	224	64457	1745	66486	117.78
11. Churu	544641	—	—	44133	—	44133	105.86
12. Dausa	194459	—	—	138146	—	138146	144.83
13. Dholpur	140557	1970	21	88295	—	90286	132.23
14. Dungarpur	108910	3920	367	6248	957	11492	111.98
15. Ganganagar	540356	466270	—	1556	266	468092	133.73
16. Hanumangarh	674786	283018	—	28489	275	311782	133.05
17. Jaipur	578605	179	—	301353	—	301532	133.44
18. Jaisalmer	77892	52994	—	15138	3	68135	116.99
19. Jalore	536769	—	—	200726	30	200756	121.32

District	Net Cropped area	Irrigation by sources					Crop Intensity
		Canal	Tank	Wells & Tube Wells	Other Sources	Total	%
20. Jhalawar	282799	3158	—	65874	556	69588	118.51
21. Jhunjhunu	325790	—	—	196443	—	196443	132.45
22. Jodhpur	190122	17	—	145254	464	145735	120.60
23. Karauli	152734	8	21	83412	43	83484	141.56
24. Kota	200065	1373	571	113049	6226	121219	129.76
25. Nagaur	737296	—	—	238021	—	238021	113.51
26. Pali	353940	—	5	62047	8	62060	107.12
27. Rajsamand	86784	—	2	4109	27	4138	103.94
28. S. Madhopur	156469	—	—	87074	76	87150	125.55
29. Sikar	422041	—	—	212096	—	212096	133.16
30. Sirohi	118186	256	826	25391	62	26535	120.09
31. Tonk	340752	—	17	68470	2582	71069	109.44
32. Udaipur	227058	398	1883	21277	78	23636	110.39
Rajasthan	**10807105**	**960105**	**7744**	**3376795**	**26954**	**4371598**	**122.30**

Source: Board of Revenue (Land Records), Rajasthan.

Area irrigated is also fluctuating from year to year as maximum cultivated area put under irrigation facilities in 1999-2000 was 5611874. Similarly the net cultivated area was highest to 21401036 hectares in 1998-99. These varied figures of cultivated area and irrigated area are the salient features of Rajasthan and reveal that even the fallow lands degraded to significant extent can yield substantial production in good water availability conditions. Though the land has been degraded to serious extent, but treatment is possible as such practices are successfully tackled in research institution but no field application could be practiced for various reasons. Even after water shortage from all the sources, cultivators capable to irrigate their lands use excessive water even the crops can be yielded successfully in low water use. The wastage of scare water is mismanagement of water resources creating various problems (Fig. 18 and 19).

Presently canal irrigated areas are seriously facing water logging problems and most of canal irrigated fields are becoming degraded for water logging problem. Silting is another problem from tank and canal irrigated system degrading the land in slow process and only cultivators of problematic areas are able to understand the degradation level. Ground water resources are depleting at faster rate and such sources are unable to coup up the increasing water requirements. Ground water level has been depleting at much faster rate during last two decades. Increasing number of tube wells has further intensified the problem and state government feels helpless to restrict cultivators to decrease excess drawl.

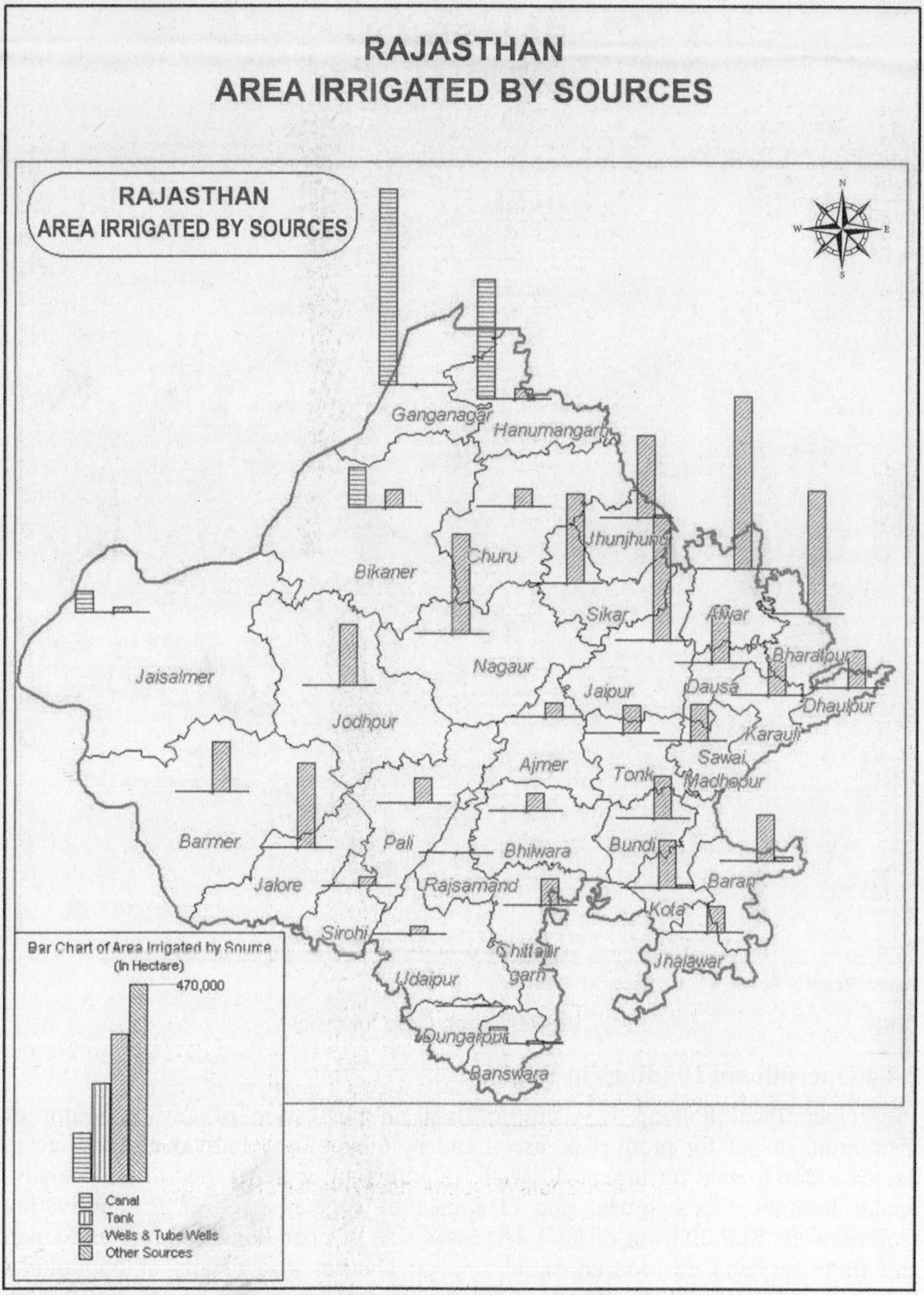

Source: Board of Revenue (Land Records), Rajasthan and Irrigation Department, Govt. of Rajasthan

Fig. 18: Rajasthan Area Irrigated by Sources

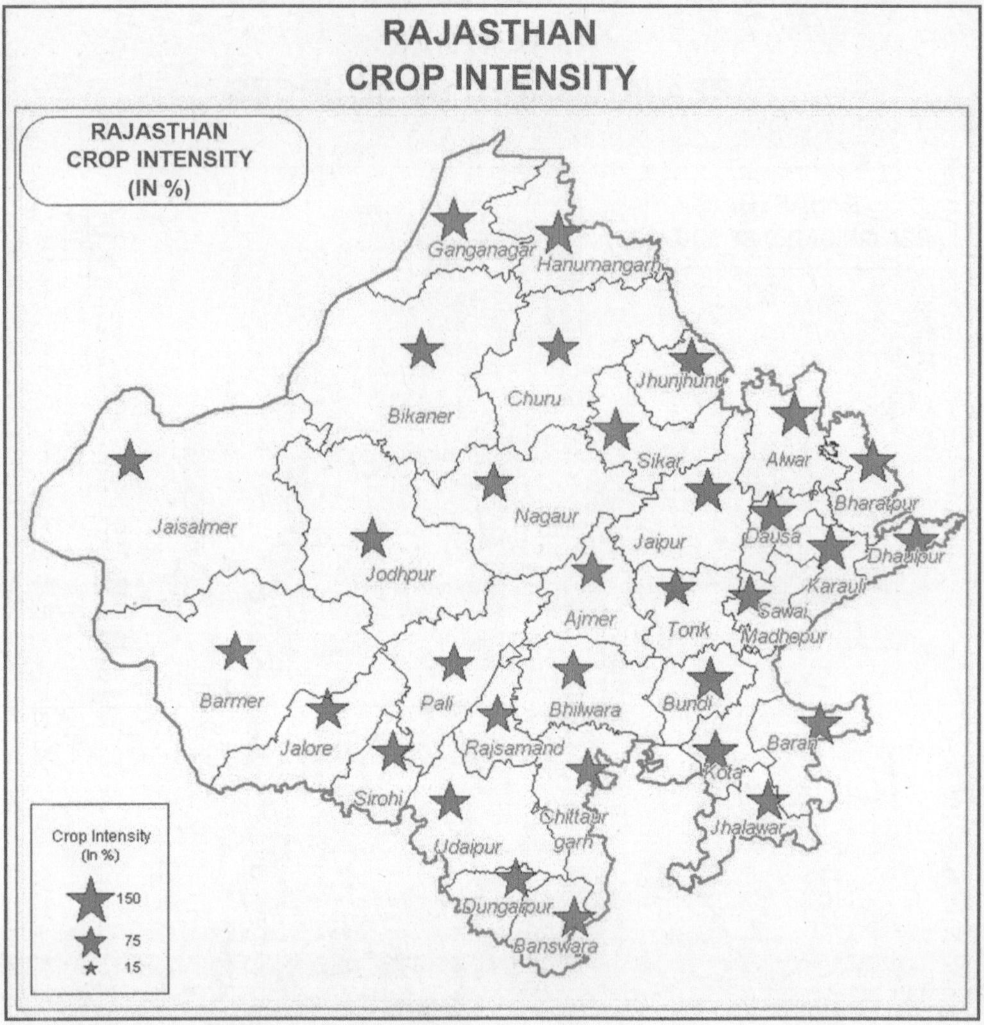

Source: Board of Revenue (Land Records), Rajasthan

Fig. 19: Rajasthan Crop Intensity

2.4.4 Operational Holdings in Region

Operational holding is a unit of land in possession of any cultivator or institution, to put for productive use. Land holding of any individual continue to decrease due to sale for urgent domestic requirement or mortgaged to any money-lender Regular encroachment and clearance of vegetative cover for cultivation requirements. Sub-division of land decreases size of operational holding and small and fragment land are consequential impact. District-wise details of operational holdings with number of cultivators of various categories and land in possession with them is given in Table 2.5 and Figure 20 and 21.

Table 2.5: Operational Holdings- Size and Number- 1995-96

(Holdings in Number/ Area in Ha.)

State/ District of Desert districts	Marginal (up to 1 Ha.)		Small (1-2Ha.)		Semi Medium (2-4 Ha.)	
	No.	Area	No.	Area	No.	Area
1	2	3	4	5	6	7
1. Ajmer	95407	40090.19	44941	65361.99	39160	109403.94
2. Alwar	136239	66639.94	73093	102522.03	52545	147224.23
3. Banswara	78457	36705.13	35804	51375.01	27089	75212.64
4. Baran	35387	18193.99	28050	41042.18	29038	82425.93
5. Barmer	7126	3420.62	9949	15267.98	24894	75152.17
6. Bharatpur	115585	53795.65	49568	71041.38	39967	112914.90
7. Bhilwara	113400	57945.24	77379	109801.22	61254	170989.26
8. Bikaner	1903	1148.45	4623	6920.77	18126	55701.36
9. Bundi	37273	19611.38	28388	41405.27	26114	73065.28
10. Chittorgarh	80079	42381.91	63982	91251.98	53269	149301.54
11. Churu	2469	1511.30	7841	12182.79	29054	87875.18
12. Dausa	34269	17165.87	25353	36093.11	22885	65051.73
13. Dholpur	55359	26479.18	23826	33993.23	16989	47444.48
14. Dungarpur	78771	31317.42	29604	42597.44	22453	61648.32
15. Ganganagar	2599	1553.97	8466	12596.97	23258	70242.85
16. Hanumangarh	9990	5886.74	18917	27776.72	37070	110473.59
17. Jaipur	74400	37375.16	59370	85938.81	64038	184476.95
18. Jaisalmer	2938	2113.87	3617	5371.16	4897	14586.08
19. Jalore	11270	6332.41	20294	30805.56	36085	103187.38

State/ District of Desert districts	Marginal (up to 1 Ha.)		Small (1-2Ha.)		Semi Medium (2-4 Ha.)	
	No.	Area	No.	Area	No.	Area
1	2	3	4	5	6	7
20. Jhalawar	48093	23989.93	34108	49287.34	32407	91739.86
21. Jhunjhunu	32746	19757.85	47283	68851.52	51218	144390.98
22. Jodhpur	15209	8457.63	22343	33814.73	43142	125323.81
23. Kota	25237	12995.64	21421	31488.34	22757	64949.65
24. Nagaur	25845	12184.10	34906	52667.28	62780	183360.20
25. Pali	56475	30373.57	47136	68767.36	45726	126185.24
26. Rajsamand	69668	29021.92	29493	41697.09	21209	59168.72
27. Sawai Madhopur	111174	50499.04	61714	87044.76	48899	139063.55
28. Sikar	40687	23283.44	50037	72401.86	58276	165398.36
29. Sirohi	27094	14542.93	19371	28078.81	15803	44492.79
30. Tonk	40352	20221.68	32434	46809.68	37466	108141.46
31. Udaipur	145168	65109.92	71702	101296.77	49210	136087.12
Rajasthan	**1610669**	**780107**	**1085013**	**1565550**	**1117078**	**3184680**

Table continued ...

State/ District of Desert districts	Medium (4-10 ha.)		Large (More than 10 Ha.)		Total	
	No.	Area	No.	Area	No.	Area
1	8	9	10	11	12	13
1. Ajmer	28163	167426.71	7740	120194.54	215411	502477.37
2. Alwar	28355	163189.72	3164	47701.38	293396	527277.30
3. Banswara	12242	69905.91	1089	18311.34	154681	251510.03
4. Baran	21728	130498.94	4162	60800.97	118365	332962.01

State/ District of Desert districts	Medium (4-10 ha.)		Large (More than 10 Ha.)		Total	
	No.	Area	No.	Area	No.	Area
1	8	9	10	11	12	13
5. Barmer	62597	424199.00	78233	1755706.36	182799	2273746.13
6. Bharatpur	23938	138911.44	2196	30215.71	231254	406879.08
7. Bhilwara	31843	185812.79	4428	66739.28	288304	591287.79
8. Bikaner	63798	434588.36	53848	1082383.50	145898	1580742.44
9. Bundi	16451	97440.11	2450	36349.85	110676	267871.89
10. Chittorgarh	32707	192736.96	4555	65974.32	234592	541646.61
11. Churu	68064	452514.69	54501	994123.61	161929	1548207.57
12. Dausa	14913	88834.97	2674	40149.12	100094	247294.80
13. Dholpur	7916	44915.67	808	11808.78	104898	164641.34
14. Dungarpur	7907	43420.07	507	11841.59	139242	190824.84
15. Ganganagar	61717	393116.65	24069	402230.67	120109	879741.11
16. Hanumangarh	56391	358999.54	23566	375256.24	145934	878392.83
17. Jaipur	46596	281863.81	12563	205150.12	256967	794804.85
18. Jaisalmer	17202	115482.00	22614	533868.31	51268	671421.42
19. Jalore	42756	273538.33	23002	391063.06	133407	804926.74
20. Jhalawar	24229	145588.15	4417	62874.31	143254	373479.59
21. Jhunjhunu	30763	179702.60	3693	50859.20	165703	463562.15
22. Jodhpur	68716	449088.35	57867	1191924.26	207277	1808608.78
23. Kota	19026	115084.17	3779	56283.59	92220	280801.39
24. Nagaur	84455	539135.73	43389	711432.67	251375	1498779.98
25. Pali	36913	229216.99	18791	351076.15	205041	805619.31

State/ District of Desert districts	Medium (4-10 ha.)		Large (More than 10 Ha.)		Total	
	No.	Area	No.	Area	No.	Area
1	8	9	10	11	12	13
26. Rajsamand	11386	66602.16	1781	28485.04	133537	224974.93
27. Sawai Madhopur	30554	180568.86	4862	72007.90	257203	529184.11
28. Sikar	42275	251652.32	7016	98243.79	198291	611279.77
29. Sirohi	11031	67750.64	3134	51132.27	76433	205997.44
30. Tonk	34118	210937.87	8986	134318.12	153356	520428.81
31. Udaipur	21739	124508.07	2797	43390.44	290616	470392.32
Rajasthan	**1064089**	**661753**	**486681**	**9101897**	**5363530**	**21249764**

Note: Karauli district information is included in Sawai Madhopur district as this district came into existence in the year of agriculture census.

Source: Agriculture Census Department, Rajasthan.

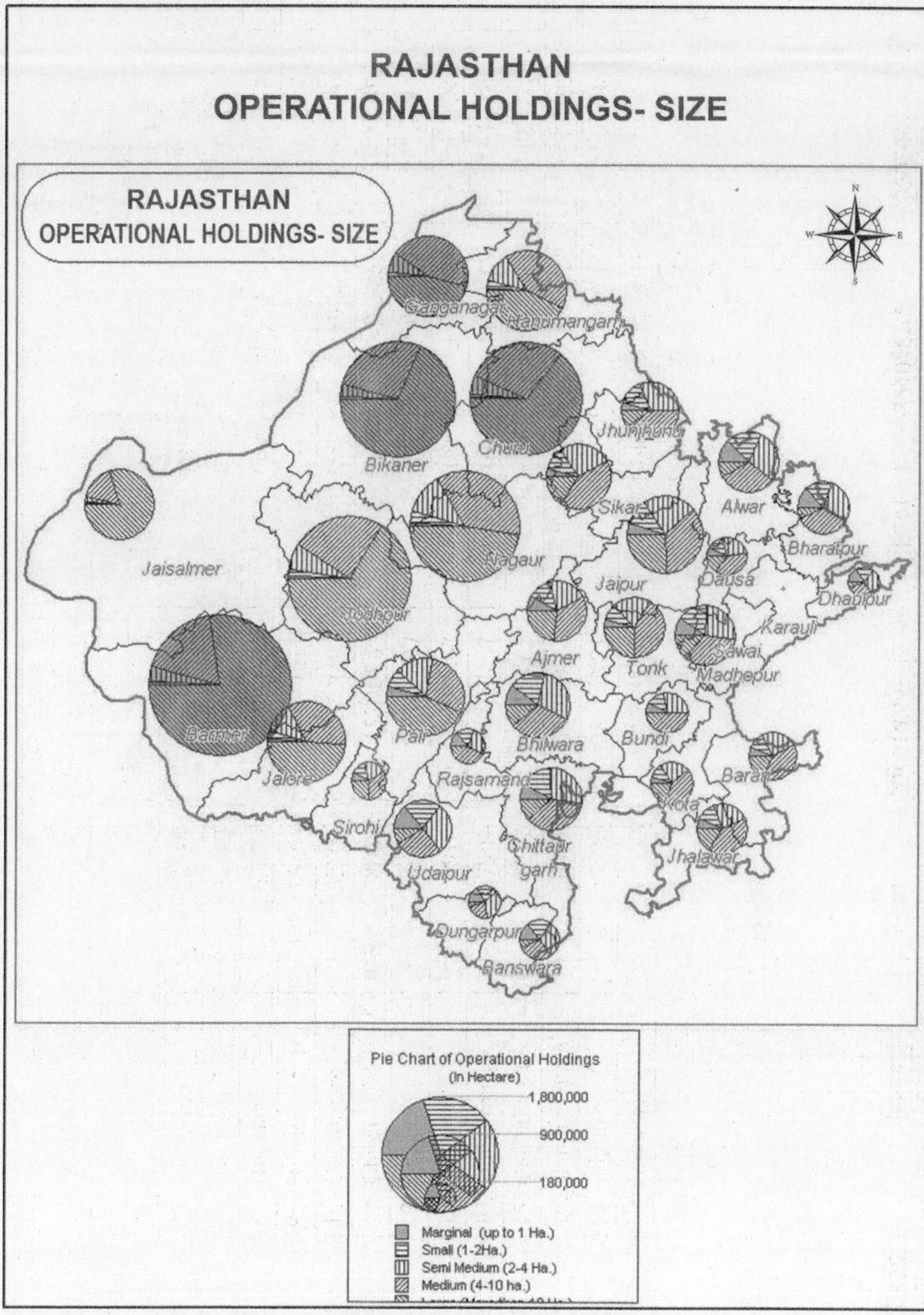

Source: Agriculture Census Department, Rajasthan

Fig. 20: Operational Holdings- Size

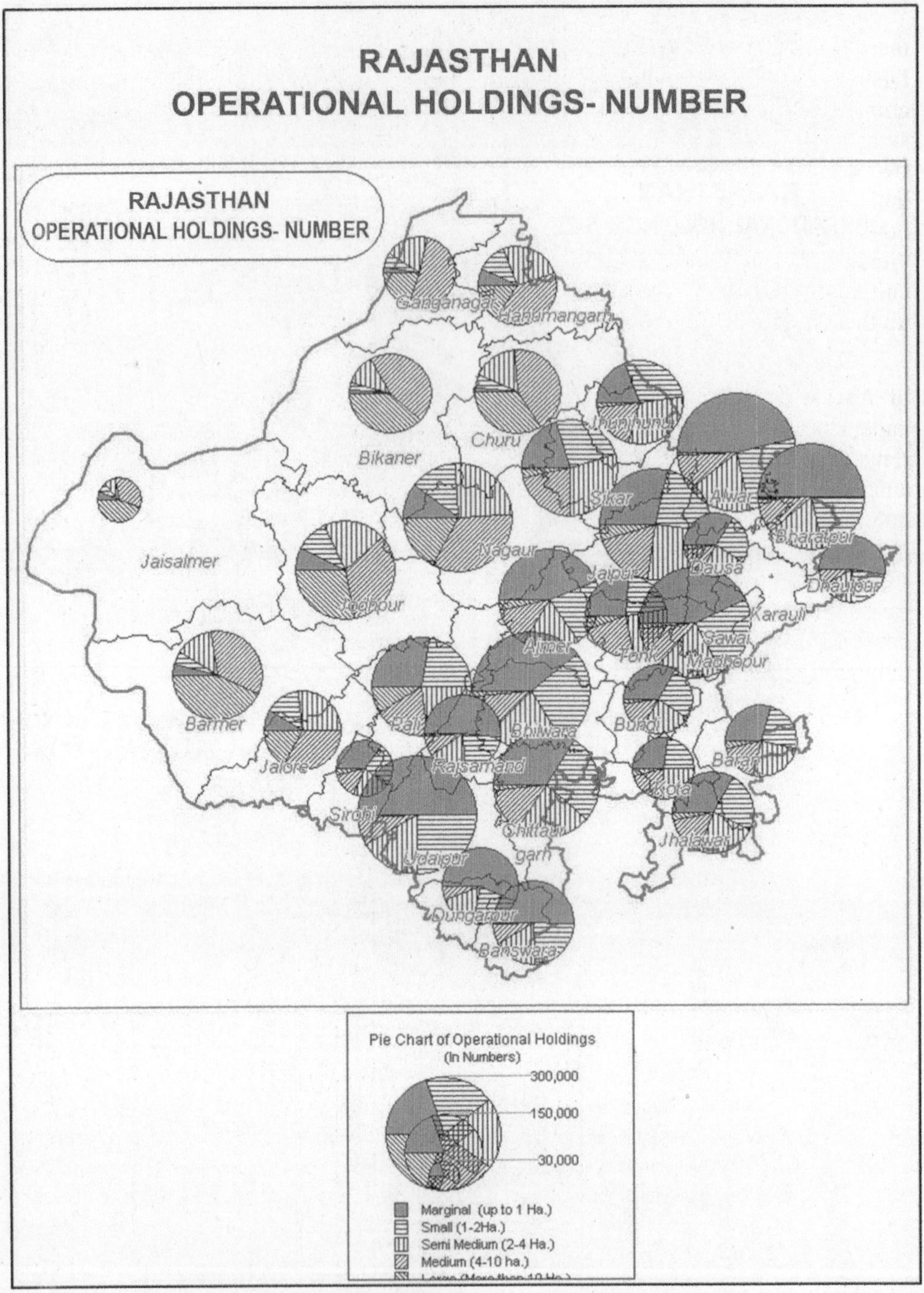

Source: Agriculture Census Department, Rajasthan

Fig. 21: Rajasthan Operational HoldingsNumbersMap

Analysis of land with ownership in terms of operational holdings reveals that there is significant variation in the size of operational holdings amongst districts. Districts of western Rajasthan have larger size of operational holdings in comparison to the cultivators of eastern and southern region. Usually marginal and small cultivators having operational holdings up to one hectare and between one to two hectare cultivable land are very poor and usually possess unirrigated cultivated land, which is not sufficient to provide their sufficient return for their livelihood. In other parts of the country where the land is irrigated and even unirrigated land can yield sufficient yield from limited land but in Rajasthan even the semi-medium cultivators having cultivable land between two to four hectares are not able to coup up their families with the agricultural production.

The total land owned by cultivators is 212.50 lakh hectares as per information of Agriculture Census Department. Similarly information as per village revenue papers reveals that total cultivated and fallow land is 207.54 lakh hectares. Thus remaining land of 4.96 lakh hectares can be under permanent pastures or tree groves and even under waste land due to severe degradation. Average size of operational holding is 3.96 hectares which is quite high in comparison to other states where pressure of population is high. Even this size of land is not feasible to provide substantial production as half of the land is marginal and seriously degraded and in case of low rainfall, cultivators deem it proper to remain unutilized as the possibilities of return is quite low. Analysis of operational holdings in terms of number of cultivators and land owned by them reveals following details:

(a) Marginal cultivators in Rajasthan state are 30.03 percent of total cultivators and cultivable land owned by them is only 3.67 percent of total cultivable land. The average size of operational holding in their possession is 0.48 hectare, which too is marginal and degraded.

(b) Small cultivators are 20.23 percent of total cultivators and land in their possession is 7.37 percent of total cultivable land. Average size of operational holding is 1.44 hectares.

(c) Marginal and small cultivators are very poor people of agrarian class and they are 50.26 percent of total cultivators of state. Cultivable land in their possession is 11.04 percent which too is mostly unirrigated and degraded.

(d) Semi-medium cultivators are 20.83 percent of total cultivators but land owned by them is 14.99 percent of total cultivable land. Condition of these cultivators is also miserable and without any other supplemental activity of income generation, they are also not able to coup up their family requirements. Average size of operational holding owned by such cultivators is 2.85 hectares, which is only economic viable in irrigated conditions.

(e) Medium cultivators owning operational land between 4-10 hectares are 19.84 percent of total cultivators who own 31.14 percent of total cultivable land. These cultivators usually able to manage irrigation facilities. Average size of operational holding is 6.22 hectares and mostly irrigated, which is substantial for their family livelihood.

(f) Large cultivators having operational holding more than 10 hectares are 9.07 percent of the total cultivators and land owned by them is 42.83 percent of the total cultivable land. These cultivators are quite well off as they can manage all the input requirements and not affected by monsoon fluctuations. Average size of operational holding of large farmers is 18.70 hectares.

(g) In case of drought conditions due to failure of monsoon, marginal and small cultivators are severely affected and need supplemental employment opportunities for their survival. If wage employment is not given in time, they are compelled to migrate to cities or other states to work as agriculture or non agricultural labour as they are unable to do any other task.

Condition of families dependent on agriculture irrespective of land size are quite miserable as agriculture sector is unable to provide even their labour cost but these people remain associated with cultivation practices as per their helplessness to opt any other activity for their survival. Position of cultivator community as a whole is not satisfactory as monsoon plays important role in agriculture yield. Out of agricultural cycle of five years, two years are usually drought years, two years are normal or sub-normal year and one year is best from monsoon point of view. Even is best production year, prices of agriculture produce are decreased to significant level and support price of the government is possible for few crops. In Rajasthan, bajra, maize and jowar are principle rain-fed crops but no substantial return is possible from these crops. Post monsoon crops are possible in irrigated land and poor cultivators have to suffer hard in case of monsoon failure.

Socio-economic conditions of half of the cultivators belonging to marginal and small farmer category is quite miserable and they need supplemental income source for meeting their bare minimum requirements. Even children of such cultivators are engaged in livestock rearing or collection of fuel wood from waste land. Their income is not sufficient to manage other activities as rural area has very limited or meagre employment opportunities. Usually such families own small livestock like goat and sheep, which graze in open causing serious environment degradation through loss of vegetative cover over the earth surface. Stall feeding of such livestock is not economically viable due to very limited earning. Presence of micro industries activities are almost negligible in the state and poor people are helpless to get additional employment opportunities.

2.4.5 Work Participation Rate

Work participation rate is one of social indicator to adjudge the socio-economic conditions of the people of the state. Work participation rate is drawn on the basis of number of days any person is able to get self or wage employment opportunities. Persons able to get employment opportunities for more than 100 days are treated as main workers and people working from one to 99 days in a year are treated as marginally employed. In Rajasthan population in age group 15 to 59 is 48.85 percent of total population. If all the people intending to work are able to get employment for more than 100 days in one year are treated as fully employed, but in Rajasthan, rural areas have very meagre employment opportunities other than agriculture sector and this sector too is unable to bear burden of total working population.

Analysis of rural work participation rate of state reveals that total 32.26 percent of total population is main workers, 13.68 percent are marginal workers and 54.06 percent are non workers. With this work participation rate of rural sector is evident that in an average family of 6 persons as per 2001 census, two persons are main workers, one person is marginal worker and three persons are non-workers. Mainly two earning persons per family are supporting their four non-earning or marginally earning members and their work participation as main workers is more than 100 days in a year. In agrarian economy, the burden on the family in general conditions is quite high and in drought or other natural calamity conditions the pressure is increased manifold.

Analysis of male population in work participation is evident that 43.70 percent of male population is main worker, 7.12 percent male are marginal workers and 49.18 percent are non workers. Position of female population is still worse as 19.98 percent of females are main workers, 13.68 percent are marginal workers and 54.06 percent female are non-workers. This fact is the crux of rural area problem as larger section of population is not able to get substantial employment opportunities even in normal conditions. District-wise details of work participation rate and category-wise distribution of working force combined for main and marginal workers is given in Table 2.6.

Table 2.6: Work Participation Rate and Categorywise Distribution

(In Percentage)

District	Work Participation Rate		Percentage to Total Workers (Main + Marginal)							
			Cultivators		Agricultural Labour		Household Industry		Other Activities	
	Male	Female	Male	Female	Male	Female	Male	Female	Male	Female
1. Ajmer	49.79	27.97	29.62	53.26	5.30	19.15	3.47	4.80	61.62	22.79
2. Alwar	53.05	43.88	55.82	70.57	6.15	12.43	2.18	2.06	35.86	14.94
3. Banswara	50.49	43.92	73.03	77.76	6.14	15.25	2.19	1.96	18.64	5.03
4. Baran	49.18	35.60	52.26	49.73	17.30	38.93	2.38	3.73	28.07	7.60
5. Barmer	51.30	41.78	66.18	83.16	3.75	5.73	2.90	2.96	27.17	8.15
6. Bharatpur	47.08	32.99	56.90	68.22	8.32	20.90	1.77	1.74	33.02	9.13
7. Bhilwara	54.42	38.61	47.92	69.49	4.32	11.65	3.15	2.90	44.61	15.95
8. Bikaner	50.17	27.52	47.29	73.16	4.92	6.99	2.77	4.12	45.02	15.74
9. Bundi	53.78	40.51	54.68	63.06	8.95	21.21	2.02	2.10	34.34	13.63
10. Chittorgarh	56.75	46.23	60.12	74.94	7.12	15.11	2.00	1.74	30.76	8.21
11. Churu	50.01	38.44	61.32	87.64	4.47	4.62	2.59	1.80	31.61	5.94
12. Dausa	45.76	36.12	57.89	79.34	3.91	10.49	2.97	2.63	35.24	7.53
13. Dholpur	51.68	33.97	56.38	35.31	6.29	9.46	1.31	3.16	36.02	52.08
14. Dungarpur	51.35	44.67	57.97	61.60	8.65	24.20	2.49	2.45	30.89	11.75
15. Ganganagar	53.55	24.96	36.87	39.68	19.17	32.41	2.06	2.54	41.90	25.38
16. Hanumangarh	51.67	29.89	54.45	70.51	14.73	18.47	2.02	2.05	28.80	8.97
17. Jaipur	47.36	22.28	26.71	59.60	2.53	9.14	4.13	5.06	66.63	26.21
18. Jaisalmer	51.73	29.37	40.82	60.13	6.25	12.34	2.30	2.63	50.63	24.91
19. Jalore	54.14	46.09	62.39	71.23	8.14	14.69	3.56	2.85	25.91	11.23

District	Work Participation Rate		Percentage to Total Workers (Main + Marginal)							
			Cultivators		Agricultural Labour		Household Industry		Other Activities	
	Male	Female	Male	Female	Male	Female	Male	Female	Male	Female
20. Jhalawar	53.90	39.55	58.82	60.48	13.73	32.52	2.07	1.60	25.38	5.39
21. Jhunjhunu	46.53	32.61	50.70	79.20	7.29	7.21	2.86	2.28	39.15	11.30
22. Jodhpur	48.43	27.09	39.34	69.15	6.64	16.38	3.25	2.48	50.77	11.99
23. Karauli	44.83	34.22	54.34	71.81	5.75	17.43	2.46	2.93	37.45	7.83
24. Kota	48.05	19.41	24.07	33.61	8.80	32.17	2.93	5.42	64.20	28.80
25. Nagaur	48.07	32.93	52.12	76.52	7.95	14.54	3.39	2.12	36.54	6.83
26. Pali	48.53	30.94	35.24	39.30	10.42	35.43	4.22	4.39	50.12	20.87
27. Rajsamand	51.41	30.02	34.02	52.75	7.52	23.52	3.27	3.59	55.18	20.14
28. S.Madhopur	47.73	35.55	55.95	75.99	5.29	13.12	2.91	3.02	35.85	7.88
29. Sikar	45.81	31.45	47.33	79.89	5.12	6.89	3.22	2.87	44.34	10.36
30. Sirohi	50.12	30.11	31.41	34.45	9.38	33.37	2.80	3.23	56.42	28.95
31. Tonk	49.28	38.27	52.03	66.34	6.19	16.82	3.49	3.93	38.29	12.92
32. Udaipur	51.49	31.83	46.83	63.03	6.04	18.89	2.56	2.45	44.58	15.64
Rajasthan	**49.95**	**33.49**	**48.06**	**67.00**	**7.18**	**16.19**	**2.85**	**2.85**	**41.91**	**13.96**

Source: Census of India, Rajasthan, 2001

Information given in above table reveals that 40.95 percent of male population is part of the state working force, which include main and marginal workers. Similarly 33.49 percent of female are main and marginal workers. The category-wise distribution of population in working force combined for main and marginal workers combined reveals that 48.06 percent of male and 67.00 percent of female workers are cultivators. Thus dependence of rural population on agriculture is quite high. Agriculture labours are 7.18 percent are male and 16.19 percent female are engaged in wage employment in agriculture sector where the payment is mainly possible on crop. In case of failure of crop, cultivators and agriculture become uncertain for their livelihood (Fig. 22 and 23).

In household industry sector, 2.85 percent male and 2.85 percent female of their respective working force are main and marginal workers. In other activities 41.91 percent of working male are engaged and 13.96 percent female are contributing their services. District specific variations persist high in work participation rate and category-wise employment position reveals that maximum 56.75 percent male are engaged as main or marginal workers in Chittorgarh district, while Karauli district has least male work participation rate of 44.63 percent. In case of female, work participation rate of 46.23 percent is highest in Chittorgarh district and Kota district has only 19.41 percent rate comprising of male and marginal workers. This variation is quite significant and main reason for high and low work participation rate is due to availability of work to the people in the respective district.

Highest and lowest proportion of male cultivators exists in Banswara and and Kota districts where the percentage of 73.03 and 24.07 respectively exists. Similarly proportion of female cultivators as main and marginal workers is highest of 79.89 percent is in Sikar district and lowest 33.91 percent in Kota district. Agricultural male workers are highest of 19.17 percent are in Ganganagar district and lowest 2.53 percent in Jaipur. Proportion of female agricultural labour of 38.93 percent exists in Baran district and lowest 4.62 in Churu district. Male workers engaged in household industrial activity is highest of 4.22 percent in Pali district and lowest of 1.31 percent in Dholpur district. Female work participation in same activity is highest 5.06 percentage in Jaipur district and lowest 1.60 percent in Jhalawar district. Work participation in other activities of male population is highest of 66.63 percent in Jaipur district and lowest of 18.64 percent in Banswara district whereas highest and lowest participation of female workers is 52.08 percent in Dholpur district and 5.39 percent in Jhalawar district respectively.

The socio-economic composition of work participation rate and activity wise association is evident for avenues of high and low participation, which is possible due to availability of the task. The main problem with rural sector is the absence of micro enterprises activities and putting excess burden over the rural economy for dependence in respective sectors.

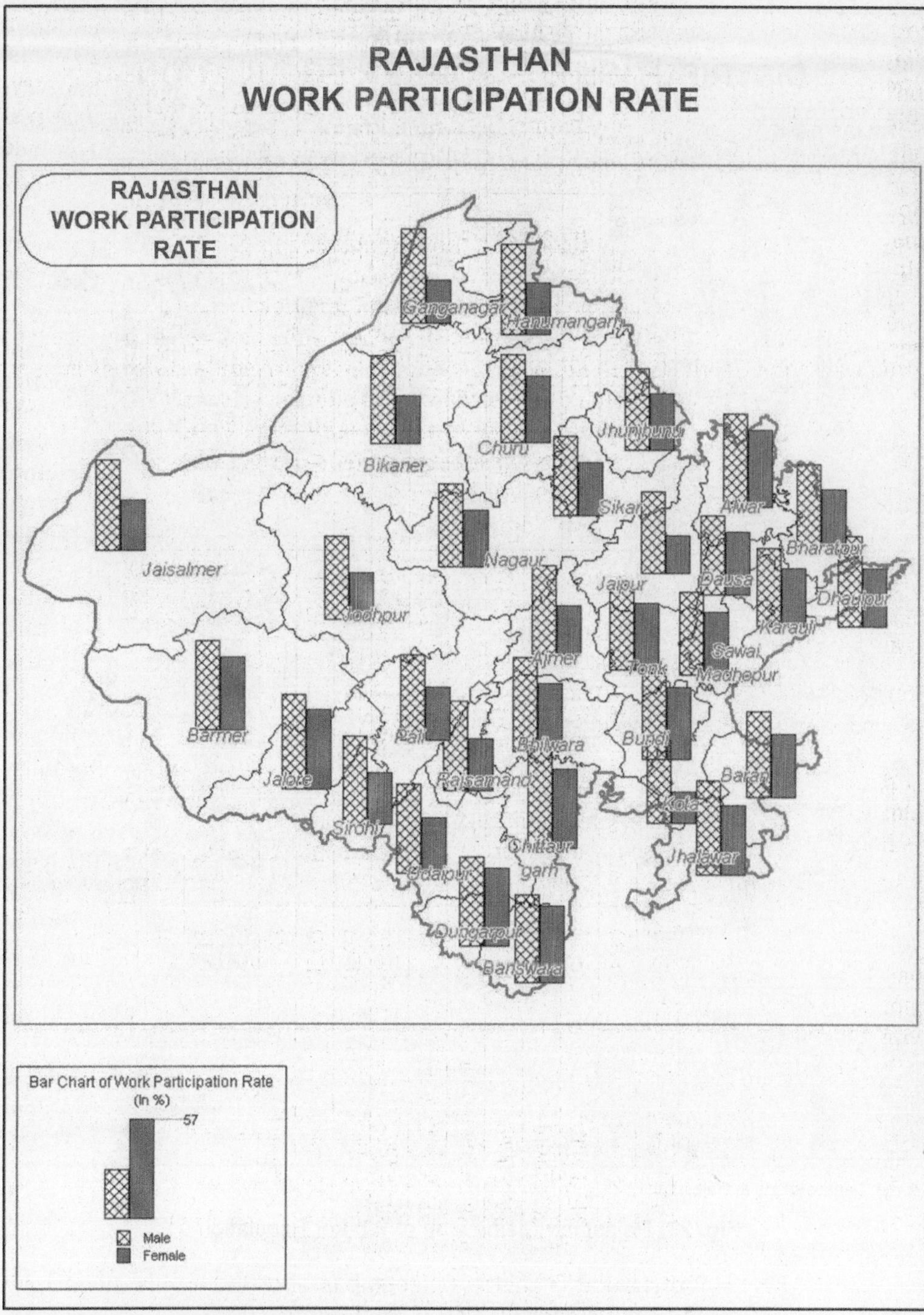

Source: Agriculture Census Department, Rajasthan and Census of India

Fig. 22: Rajasthan Work Participation Rate

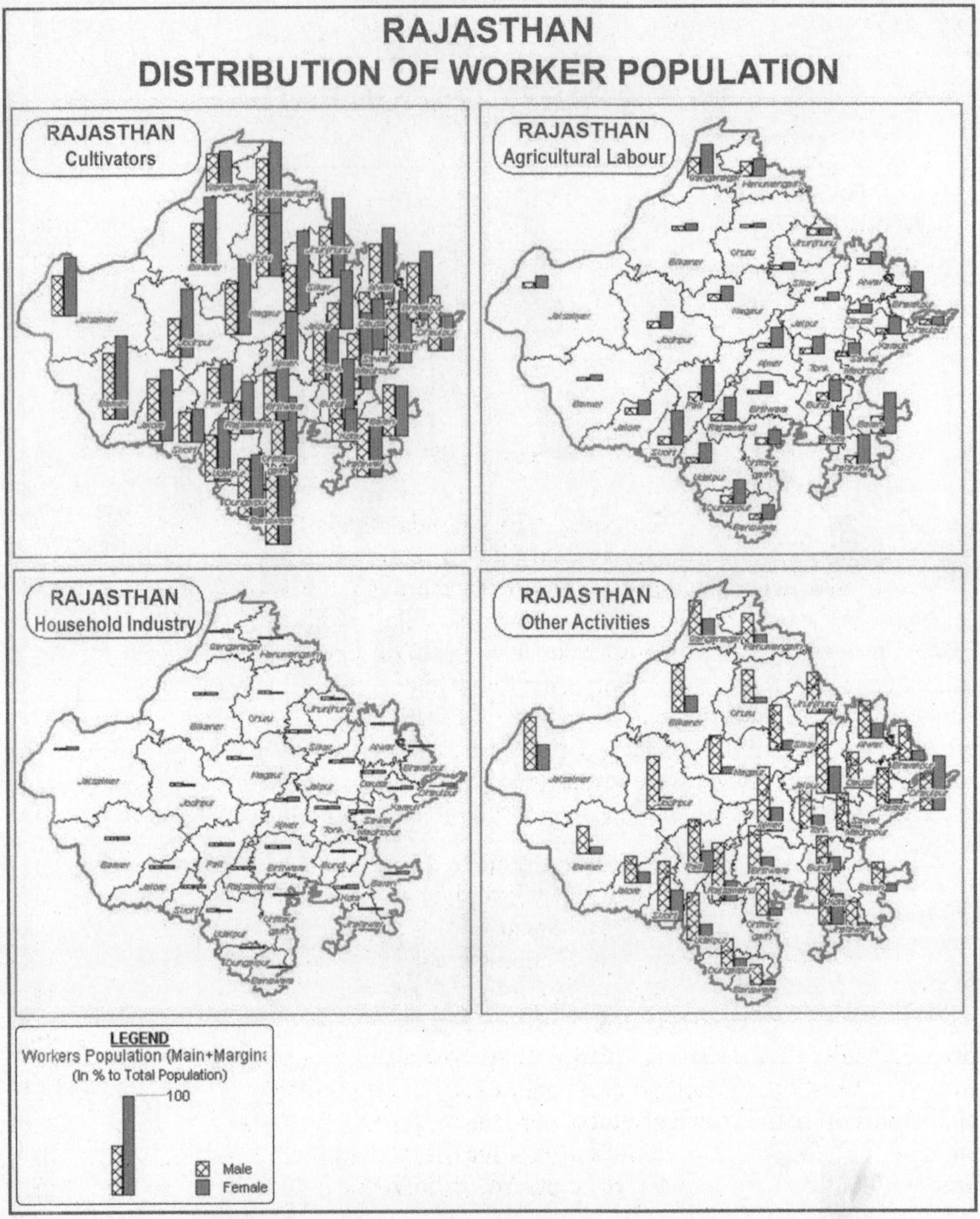

Source: Census of India, Rajasthan

Fig. 23: Rajasthan Distribution of Worker Population

3

Perception and Recollection of Drought

Contribution of agriculture including animal husbandry ranges between 22 to 33 percent of state net domestic product at current prices. Agriculture is dependent on rainfall and there is significant fluctuation in the agriculture production varying between Rs.15451.83 crores in 2002-03 to Rs. 21943.45 crores in 1998-99 with proportional contribution of 20.59 to 33.49 percent. Average contribution of agriculture sector In average terms the contribution remained Rs. 19293.31 crores. In average percentage terms the contribution of agriculture including animal husbandry remains at about 27.04 percent of state domestic product. The estimated contribution of agriculture sector alone is about 15 percent of net state domestic product. Agriculture sectors in Rajasthan state suffers from rainfall problem to vital extent and level of fluctuation is about more than 7 percent on both extremes from average position.

In developed economy, the growth of agriculture sector remains at about four percent per annum and its contribution to net domestic products does not exceed from 3 to five percent as growth of other sectors becomes as high as 40 percent per annum. Rajasthan state is stagnant in view of agriculture sector as the production and productivity remains fluctuating and any good rainfall year provides highest production, which is unable to be achieved in coming five to seven years. The significant fluctuation is the result of erratic monsoon which disrupts total agriculture progress of the state and it is not possible to overcome the problem in view of uncertainty of monsoon.

Monsoon plays vital role in the development of the state and state economy suffers from this setback to great extent as effect of agriculture production reflects on state economy and net domestic products decreases in bad monsoon year. Monsoon and agriculture practices are inversely correlated as good monsoon helps in better crop prospects and monsoon failure is responsible for crop failure. The area spread and good or worse impact can be for total state or may affect any region. The monsoon is adjudged with the arrival in time and regular spells of rains as per crop requirements, but this is utopian condition which is far away from the practical situation. Occurrence of monsoon is favourable for crop if it receives in time i.e., during last week of June or first week of July. Further spells are expected with intervals of a fortnight and such three spells are received to meet the crop requirements, the crop yield is bumper. However, such condition do occur once in five years and also with varied nature.

In district like Jaisalmer, where annual average rainfall is 18.55 cms. minimum rainfall of 4.40 cms occurred in 2002 depicts the worst situation the local people had to face as consequential impact of the adversities. Normal annual rainfall of Rajasthan is 57.51 cms but average rainfall during 2002 remained 26.71 cms, which was worst year for the state. In Banswara district where highest rainfall occurs and normal annual rainfall is 95.03 cms, it remained lowest of 52.46 cms in 2002. Thus, the situation varies from district to district and any deviation in occurrence of rainfall creates hardship irrespective of the fact that it is better than other district.

The yearly total rainfall is highly variable in different regions of the state and it is most erratic in eastern half with frequent spells of drought, punctuated occasionally by heavy downpour in some years associated with the passing low pressure systems over the region. The south-west monsoon which has its beginning in the last week of June in the eastern parts, may last till mid-September. Pre-monsoon showers begin towards in the middle of June and post monsoon rains occasionally occur in October. In the winter season also, there is sometimes, little rainfall associated with the passing western distribution over the region. At most places, the highest normal monthly rainfall is during July and August. The number of rainy days during this period varies widely in different places ranging from 10 days in Jaisalmer to 40 days in Jhalawar and 48 days in Mount Abu. Rainfall during rest of the period ranges from 2.1 cm at Jaisalmer to 7.2 cm at Jaipur distributed over 2.5 to 6 rainy days.

3.1 NATURE OF DROUGHT PERCEPTION

Drought is universal phenomenon, which almost depends entirely on the parameter of water. Though surface and ground water resources are there to supplement the water requirements, but these perennial resources of water are also dependent on rainfall, received over the earth surface. Rainwater is charged to these resources and can help in meeting the annual requirements of all the sectors where water is essential. The total quantum of rainfall and its spatial as well as temporal distribution received by an area determines the severity of drought. It is always concerned with the inadequacy of water either for irrigation or drinking purposes. In general, the scarcity of water which does not fulfill the requirement of particular object creates drought conditions.

An area, which receives normal rainfall but after passing the harvesting season, it does not yield for proper crop yield, it signifies that agriculture could not yield in the requisite manner, it is worst for agriculture and affects most to the people associated with cultivation practices. Similarly, even when the normal rainfall is unable to yield much for meeting the drinking water requirements for the people and livestock, it is more severe than agriculture drought. Similarly, any area receiving rains below the normal rains, it affects agriculture and drinking water requirement, the impact is more severe as all the sections of the society are severely affected. Thus quantum of rainfall and time of receiving is the major factor, which has severe impact over any area. The area spread with the problem and its intensity are the major factors which deserves timely attention for alternative measures to tackle the related problems.

3.1.1 Definition of Drought

There is no universally accepted definition of droughts. Early geographers defined drought as a prolonged period without rainfall. The Meteorological Office (1936) defined absolute drought as a period as a period of at least 15 consecutive days none of which had rainfall of 0.25 mm or more. According to Ramdas (1960), drought was taken as a situation when the actual seasonal rainfall is deficient by more than twice the mean deviation. In USSR, drought is defined as a period of 10 days with a total rainfall not exceeding5 mm. Gibbs and Mahar (1967) made a study of droughts in Australia by using annual rainfall deciles as drought indicators. Using a network of 100 stations, they draw isolines of annual rainfall deciles and took areas covered by the first docile range as drought areas. Using similar criterion, droughts in Bihar studied by George and Kalyanasundaram (1969). Van Roy (1965) developed a drought analogy index based on rainfall departure and mean of lowest ten values of rainfall in series.

A slightly modified criterion was adopted by Kalyanasundaram and Ramashastry (1969) for study of droughts in Bihar state. According to Bates, drought occurs when the annual precipitation is 75 percent of the normal or when the monthly precipitation is 60 percent of the normal. The Meteorological Department of Government of India, annual rainfall of 75 percent or less of normal is taken as a drought year and 50 percent or less as a severe drought year.

Drought is, however, evaluated in a more meaningful manner balance computation taking in to account the rainfall pattern, evapo-transpiration, moisture characteristics of soil, resting depth and stage of crop growth etc. Van Bavel (1953) found that the amount of moisture deficit is quite sensitive to change in moisture storage capacity of the soil. According to his study, the average number of drought days decreased from 45 to 10 as the moisture storage capacity increased from 25 to 100 mm.

In semi-arid zones like Thar, the specification of drought cannot be divorced from the land utilization. Rainfall too meagre to produce crops may be adequate for pasture. An amount of rain, which would give normal growth of indigenous pasture, may prove inadequate for another area. Thus climatic criteria for drought varies with the type of crop growth, pasture species and the soil type of the locality. The effect of the continued drought in a pasture land would be deterioration of pasture, either by failure of plants to germinate or grow or by reduction of soil moisture to such an extent that even perennial plants cannot maintain themselves. Hence, in the study of the droughts in pasture land, concept of initial effective rainfall is necessary to initiate plant growth following drought and an effective carry over rainfall to maintain growth periods by water balance models during individual years are required. Such analysis would indicate duration of non-growth or drought periods within the growing season (Fitzpatrick *et al.*, 1967, Krishnan, 1969).

The Directorate of Commonwealth Bureau of Meteorology in 1965 suggested a broad definition of drought as "Severe Water Shortage." It means shortage of water in comparison to the needs of drinking water for habitations and animals in a particular area. The shortage of water affects three main users of water i.e., plants, man and animals. The intensity of water shortage is often reflected in the migration of animals during severe drought years and also effects the agriculture and to some extent industrial production where water use is essential. It causes miseries in rural people, whereas the urban folks, insulted by the industrial production and supply of tap water, drought is sometimes known only through newspapers. But for rural habitations, drought is personally identified and its effects are often practically experienced. The communities mostly dependent on primary production system, face the consequences of drought while the community depending on secondary or tertiary occupations rarely confronted with the impact and severity of drought.

Definition of drought given by Folly is more elaborative, which treats drought as "a period of rainfall deficiency extended over months or years of such nature that crops and pasturage for livestock are seriously affected, if completely burnt or destroyed, water supply resources are seriously depleted or dried up and sheep and cattle perish." In Britain, where drought was defined more specifically in 1887 as, "fifteen consecutive days, each with less than one hundredth of an inch of rain, the basin was and still urban water shortage capacities to town water supplies." This definition is based on the specific problems of the country and its universal impacts are not fully covered.

Report of the Second Irrigation Commission in 1972 states, "Drought is the result of an imbalance between the soil moisture and evapo-transpiration needs for an area over a fairly long period so as to cause damage to standing crops and a reduction in crop yield." The basic characteristics of drought is steady rise in temperature in addition to the absence or severe deficiency of rainfall over a long period. Several factors such as precipitation, temperature, wind velocity, sunshine, soil texture, soil moisture and antecedent rainfall interact to produce the situation. The key role is played by rainfall and the crucial variables are its distribution, its variability and its capacity to meet evapo-transpiration needs.

The Indian Meteorology Department had defined drought where annual rainfall is less than 76 percent of normal. Moderate drought is stated to prevail where rainfall deficit is between 25-30 percent and severe drought when the deficiency is more than 50 percent. Areas, where such deficiency occurs in more than 50 percent of the normal rainfall, are considered as drought areas. Areas where such deficiency occurs in more than 75 percent, are considered as cronic drought areas. This definition is elaborative but does not consider the time spell of rainfall, which is a critical factor.

The definitions of drought are not exclusive to cover all the related factors as there are various geographical, environmental, social and economic factors which are vital as the drought is very severe disaster. In this regard, National Commission on Agriculture has defined drought in 1976 as occasion when the rainfall for a given week is half of the normal or less, when the weekly rainfall itself is 5 mm or more and agricultural drought has been defined as a period of four such consecutive weeks in the period from middle May to middle of October or six such consecutive weeks during the rest of the year. Here quantification of rainfall is a problem as areas receiving less than 20 cm and more than 200 cm cannot be identical on single parameter. There is necessity to define all the situations in relative context for addressing spatial variations.

Drought is a disaster, which is known in advance and people need to carry out requisite preparation to face the consequences. Any definition covering all the causes and consequences keeping in view the intensity and impact is the best to be followed and used. Thus drought is a temporary reduction in water or moisture availability significantly below the normal or expected amount for a specified period. The impact of drought results into shortage of food, fodder, fuel wood and water or in discrepancies between supply and demand for food, fodder, fuel wood and water, which are primary needs of the rural masses.

3.1.2 Role of Rainfall

Low rainfall coupled with erratic behaviour of the monsoon in the state makes the state most vulnerable to drought. Drought conditions are the resultant impact of low rainfall which have number of eventualities for low and erratic conditions. Occurrence of rainfall in time is the sign of hope for substantial rains to meet all the human requirements. It is true that agriculture in Rajasthan is based on monsoon as most of the cultivation practices are rain-fed. Even in areas of irrigation facilities, monsoon plays vital role. Cultivation practices are started after first rain and three spells of fortnight variation are best for cultivation of crops. When first spell of monsoon is delayed by a fortnight, cultivators try to sow crops which can yield in short duration. After first spell of rains, sowing activities are done in the fields and germination process starts, which require water for retention of plants and growth. If the second spell of rains is delayed by three weeks or more from fist spell, newly grown plants are extinct and crop failure is most critical. If second spell is in time and third spell is delayed, there is possibility of some fodder for livestock. When plants are in the process of formation of grain, rainfall helps in strengthening of the crops and hope of good crop is felt by cultivators. Thus total cultivation process is closely associated with monsoon behaviour and any deviation of vital nature results into crop failure. Thus the cultivator remains at stake at each spell of rainfall.

Rainfall plays vital role in the life of rural people who are directly affected with erratic nature of monsoon. Cultivation is directly affected as this is the principal source of livelihood of most of the rural people. Rainfall in Rajasthan is lifeline of the people as in addition to human population, livestock is also sufferer. Drinking water problem becomes most critical in case of monsoon failure. Monsoon in natural factor and it plays its role as per the environmental and geographical conditions, human interface is also responsible for erratic nature of monsoon as loss of vegetation is directly related with good rainfall. Frequency of drought in Rajasthan as per past analysis is given in Table. 3.1.

Table 3.1: Frequency of Droughts in Rajasthan

S.No	Recurrence Period	Districts
1	Once in 3 years	Barmer, Jaisalmer, Jalore, Jodhpur and Sirohi
2.	Once in 4 years	Ajmer, Bikaner, Bundi, Dungarpur, Ganganagar, Hanumangarh, Nagaur and Churu
3.	Once in 5 years	Alwar, Banswara, Bhilwara, Jaipur, Jhunjhunu, Pali, Sawai Madhopur, Sikar, Dausa and Karauli
4.	Once in 6 years	Chittorgarh, Jhalawar, Kota, Udaipur, Tonk, Rajsamand and Baran
5.	Once in 8 years	Bharatpur and Dholpur

Source: Drought Management and Relief Department, Rajasthan.

Situations narrated about recurrence of drought are based on trend of last one hundred years. Out of this duration, there not more ten years when no area of the state is gripped by drought. The past trends reveal that drought remains in the state in any part or region. However, there are situations when the drought is spread over total state as in case of 2002 or there are conditions when some part of a district is affected by drought. Therefore, drought is serious when total state is affected by drought and number of villages and population affected is more than 75 percent of rural areas. Rajasthan is the main sufferer of drought as there is dryness prevails in most part of the state and drought affects most in hyper-arid, arid and semi arid regions of the state.

3.1.3 Difference between Drought and Famine

Generally people use drought and famine treating it a common phenomenon but there is significant difference in drought and famine. Drought is natural condition caused due to failure of monsoon on any reason. This is related with the occurrence of rainfall, which is erratic and uncertain. Drought is a situation when the monsoon is significantly below than the normal rainfall. It has relationship with the climatic impact and there are forecasts of low rainfall prior to its occurrence. Every year Meteorological Department forecasts about the behaviour of monsoon in general terms. These forecasts are generally based on the temperature and pressure of air over the surrounding areas of seas, which provide monsoon in the country. Indian monsoon occurs from Arabian Sea and Bay of Bengal and forecasts are issued after thorough assessment of the behaviour of these regions.

The forecasts are generally falsified as there are various other factors which cannot be adjudged in advance and main feature is the wind velocity which helps in setting the direction of clouds. These winds usually change quite frequently and forecasts issued are generally not maintained in its version. It is true that clouds generated from seas move to any direction with the support of winds and there is every possibility of change in the direction but rains do occur in some parts or other. Therefore, the forecasts can be true for country as a whole but regional variations are possible as wind directions regularly play vital role in setting the directions of the clouds.

In case of Rajasthan, Aravalli mountain ranges play significant role and western Rajasthan usually receive less rains in comparison to eastern and southern regions of the state. In western desert region, sand dust play negative due to wind storms, which hampers in setting the wind direction at this region. One of main constraint of low rains in desert regions of western Rajasthan is loose sandy dust, caused due to large scale felling of trees from this region. Occurrence of rains in total Rajasthan has been decreased significantly in past forty years, which is responsible for drought conditions.

Famine is the resultant impact of drought and basically social and economic factor. In severe drought conditions, crop failure to significant extent creates various problems like food shortage, non-availability of fodder as agricultural residue and loss of purchasing power of rural masses. This situation is devastation as survival of human life, livestock and vegetation is seriously affected. There is acute water shortage and people leave their livestock as drinking water is not sufficient for their own needs. Between 1997 and 2003, livestock census reveals ten percent decrease as resultant impact of 2002 famine, which spread in all the 32 districts of the state and livestock loss was severe due to non-availability of requisite fodder and water.

Therefore drought is natural factor and famine is the social and economic factor, which the rural people suffer due to loss of crops, fodder, vegetation and water. The severity of drought faced by people and meagre condition is the consequence when it has devastating impact. Usually people of Rajasthan have their capacity to face drought to great extent, but famine conditions put loss for their survival and existence. Development of Rajasthan state suffer vitally due to famine conditions.

3.2 PEOPLES PERCEPTION OF DROUGHT

Usually rural people of Rajasthan are well versed with drought due to its recurring feature. There are examples when the state suffered with consecutive drought conditions. Rural people have their own conventional systems for assessment of rainfall even one or two months in advance based on their traditional practices and their own predictions are near correct in view of the knowledge transmitted from their ancestors. People start their planning for facing the drought conditions on the basis of their own assumptions and when such forecasts match with government presumptions, they become sure of the monsoon failure.

Drought assessment is made by the cultivators on the basis of first monsoon rains in terms of occurrence in time or delayed. Cultivators carry out their preparations in advance through ploughing of their fields and removal of seeds In case of low intensity of rains and disturbance in further timely rain spells, cultivators make their assessment of crop failure. This is the first hand information source about drought and newspapers spread this news to draw attention of the government for making their own assessment of crop damage assessment.

Rural people are able to assess the drought situation when the government machinery is alerted for making assessment of crop damage in all the districts through their own system. Cultivators start their own preparations in view of their ability to face drought conditions. If further spells of monsoon remain erratic or uncertain, cultivators are able to make their personal preparedness of facing the drought conditions as they are well versed with their capacity to face drought. Most sufferers of crop damage are marginal and small cultivators, who are 50.26 percent of total cultivators owning 11.04 percent cultivable land. In all the drought conditions, these people remain main sufferers of drought conditions, who are compelled to take appropriate decision to stay at home and wait for government preparations to provide wage employment but others migrate to other areas in search of employment.

Semi-medium, medium and large cultivators are also affected to great extent as surface and sub-surface water is also affected with low rainfall conditions. In case of wide spread monsoon failure, total cultivation practices based on rainfall are seriously affected. Marginal and small cultivators are well aware of their low cultivated land and employment is possible to 3 or 4 months. Thus such people keep their supplemental income source to manage their livelihood as single crop in very meagre land cannot manage their family requirements even at the bare minimum level of survival. People with low cultivated land are most hard struck people but they migrate to cities in search of employment. Livestock rearing is the supplemental income support specially in drought conditions. Such people keep small livestock like sheep and goat which is accustomed of open grazing. Generally people of Rajasthan face serious problem due to recurrence of drought and their perception about drought is multi-faceted.

3.2.1 Source of Alternative livelihood

Drought is recurrent feature of the state and agriculture alone is not able to suffice requirement of the people at large. Generally, more than 60 percent of cultivated land is unirrigated which is sole dependent on monsoon. This provides employment for 3 to 4 months in year and people want some supplemental source of earning to meet their bare minimum requirements. If alternative employment opportunities are made available to them, impact of drought would be subsidized to great extent. Even people dependent on agriculture are 50 percent of total state population and adding agriculture labour number the problem for supplemental employment is quite a rigorous task but this is essential to get rid from the hardships caused to most of the people.

Most of the people of western Rajasthan, who suffer most from recurrent droughts, possess large number of livestock, which helps them to significant extent, but people want some micro-enterprise activities based on local resources. Even livestock require marketing linkages which are possible with establishment of dairy plants for collection of milk. Dairy provides fortnightly payments which can be helpful in meeting their domestic requirements. Rural unemployment position is quite serious as micro industrial activities are completely missing in remote areas and people are not properly skilled to manage such activities. If sources of alternative employment opportunities are created in most of the rural areas, intensity of drought can be minimized to great extent.

People are aware that government helps the people through wage employment in drought conditions. If such measures are carried out on regular basis for creation of permanent assets, these can help the people in facing drought impacts. Employment generation in drought prone areas is the necessity of people. Rural areas have huge potential for creation of permanent assets which can be helpful on sustainable basis. Condition of marginal and small cultivators is almost equal to agricultural labour, who want assured supplemental income generation activities for their assured livelihood. Employment generation in rural areas can be helpful to great extent and rural people can equate them with urban employed people, who have sufficient opportunities.

3.2.2 Treatment of Degraded Lands

People reveal that cultivated lands are mostly degraded and need proper treatment. If productivity of land is increased by treatment of land, their drought facing capacity in increased to significant extent. Usually cultivators are vigilant of their degraded lands but unable to treat in view of lack of resources. They are also not aware about the technology for treatment of specific degradation problems. Causes of land degradation can be known with soil and water tests of any area, the test results can also supported with proper advice about the mode of treatment, which the technical experts can very well communicate. If degraded and marginal lands are treated with the infested problems, cultivation practices can be purposeful to them. People also feel that water-harvesting measures can be helpful to vital extent.

People now started realizing that drought is not only natural factor but clearance of vegetated land for meeting cultivation requirements is the ultimate cause of failure of monsoon in hyper-arid, arid and semi-arid regions. The vegetated land used for cultivation purposes has lost its moisture and most of the cultivated land is in the process of losing its productivity. But one person cannot change total degradation of earth surface. People continue to cultivate in marginal lands which ultimately become barren. Presently fallow lands are about equal to cultivated land and operational holdings own by people are unutilized for the reason that cultivation in such lands is not economical. This problem is the helplessness of the people in view of growth of population at tremendous high rate.

3.2.3 Treatment of Surrounding Lands

People are aware of the dryness in the environment causing serious problems in cultivation practices. Present efforts for sand dune stabilization and use of wasteland and barren land for economic use are not effective and need concerted efforts. If surrounding area of cultivated fields are vegetated, the moisture retention in cultivated land can help their cultivation practices in case of erratic or low monsoon conditions. Vegetation in the surrounding area can be helpful for check of dry atmosphere and also helpful in low evaporation of water or moisture on surface and sub-surface levels.

People of rural areas not feel importance of vegetation in cultivation practices. Usually trees grown in forest lands, community lands, waste land and barren land are cut for meeting fuel, fodder and timber requirements. Such acts are performed by few powerful and influential persons. Usually people are well aware with the importance of trees, but they are unable to check cutting of green trees. Droughts have compelled people to cut trees in scarcity condition to survive their livestock as no fodder is available and survival of livestock was their necessity.

People in general feel the denudation of land is harmful for atmosphere but unable to check the acts of few persons, who carry out such practices even after knowing the consequences. People themselves facing the consequences of human interface in the form of clearance of vegetation, felling of trees and soil erosion as impact of their own acts. People in rural areas used to retain some land for large scale plantation called orans, which were associated with their ethnic values. Cutting of trees or branches was not permissible even to local residents. But such values are diminishing and such orans are losing their importance. Wind storms flow uninterrupted creating various environmental problems but people are helpless as people are least interested to grow plantation in their surrounding areas.

3.2.4 Credit Support

Banks and other credit institutions have directions to provide loans to people up to Rs.15000 without any mortgage or guarantee, but banks hesitate to provide loans to rural people in view of recovery problems. Rural people of poor economic status want credit support to face the drought conditions and create additional employment opportunities specially self employment with such help. In drought conditions, poor people face serious problem in availability of credit in hard conditions and moneylenders, who provide loan in bare necessity charge heavy interest. Even in case of death, serious diseases or religious rituals, their credit requirements are met by moneylenders. The poor people are exploited to great extent as government support for bank loan is completely ineffective and poor people face serious problems.

3.3 PERCEPTION OF DROUGHT PROBLEMS

Impact of drought is very serious as the total environment of Rajasthan is seriously disturbed. Every drought is dangerous for earth surface as environment degradation is serious with increased intensity. Usually wild plants grown during rainy season are vanished by stray livestock and soil structure is trampled with frequent movement of grazing animals. In case of drought conditions, starting famine relief works has become customary to provide wage employment to needy people for their survival. Such employment oriented activities are not associated with permanent creation of assets but digging of pits to put on roads, which is cleared in rainy season or blown by winds.

Survival of livestock is also felt need and fodder depots are established to provide fodder carrying from other parts of the surrounding states. Generally such arrangements are made for cattle but stray cattle are in large number and in normal conditions, those are grazed in open area. No stall feeding of cattle in\s in practice but fodder arrangements are made to express the faith of government for survival of cattle. Stray cattle are left in open or carried to cattle centers for survival during the drought period, which further stray in open as per usual practice. Small livestock like goat and sheep are migratory by nature and people carry in herd to surrounding states as per their usual practice. In drought duration camel are also become migratory as land surface is completely denuded by grazing livestock.

Ground water depletion is serious problem of Rajasthan state as most of the irrigation and drinking water requirements are met with ground water resources. The ground water resources are very limited in view of low and erratic rainfall and only 20 percent of rainwater reaching over the earth surface is charged to ground water resources. Chances of self charging of ground water resources are very limited. Tube well digging has further aggravated the problem as such units draw maximum water from sub-surface. In view of constant over drawl of water, ground water depletion has reached at the critical level where further depletion is dangerous. Almost total state is gripped with serious ground water depletion problem.

Ground water surveys undertaken by Central Ground Water Board reveal that most of the areas have been converted into dark zone due to excess drawl of water for meeting irrigation and drinking water requirements. Few decades back, rain water harvesting was practices in most parts of the state as people were accustomed to save water and use rain water for their most of the requirements. With increase in population and creation of drinking water arrangement in most of the rural areas, traditional practices of rain water harvesting are almost vanished as people are able to get water at their homes or in villages. Now ground water resources are at the verge of extinction and people continue to dug their tube wells at further depth to get water for irrigation and drinking water needs.

In summer months many villages are provided water through tankers and this system is prevail in case of monsoon failure. Drinking water requirements are associated with human facility and increasing population has disturbed total system of utilization of ground water resources. It has been visualized that dams constructed for irrigation requirements have become drinking water reservoirs in view of meeting the requirements of urban population. The norms fixed for drinking water arrangements for urban areas are 100 litres per capita day and for rural areas

40 litres per capita day The rural population keeps livestock and the variation in water supply norms to urban and rural population is distinguished. Furthermore, in most of water supply arraignments, clean water is provided for all human and cattle requirements, which most of human and livestock needs can be met by recycled water. Irrigation of crops is necessary for meeting growing population requirements and there is necessity for proper coordination as all the water requirements are creating serious problems.

Water crisis is being felt by the people and generally people habitated near water bodies protest for carrying water to distant places by sacrificing their requirements. This problem would create serious problem in coming years as water resources are diminishing due to mismanagement. People in possession of water bodies try to use maximum without bothering for the requirement of other people. This is generally realized in ground water resources use as every person want to dug further to snatch maximum water. There is no consideration of diminishing water resources but every one is interested for present use without considering the consequences in coming years. Ground water resources are reached at significant depth and use of such water is degrading land through salinity and alkalinity. This process is responsible for decreasing productivity but no one is bothered for the position in near future.

In case of surface irrigation, cultivators of front end to resources try to use maximum water to irrigated their crops while the cultivators at tail end suffer for non-availability of water. Constant use of excess water in the fields has created water logging problem in most of the surface irrigation fields. Water logging is serious degradation of land where water level starts coming up and in due course of time, the cultivated fields are serious affected from seepage water and productivity of land is decreased significantly. Cultivated lands of Ganganagar and Hanumangarh district in vicinity of canal and irrigated with canal water are seriously affected water logging problem. Cultivated fields of Jaisalmer and Bikaner would become problematic after five years. There is danger of severe food crisis in view of water logging problem.

Use of chemical fertilizers and chemical pesticides is increasing tremendously without understanding the consequences on human body and environment. Various birds and animal species are extinct due to ill effect and half of human diseases are caused due to use of these chemicals. In water logged areas cultivators of canal areas are using chemical fertilizers and pesticides almost double the permissible limit, which is producing poisonous agriculture produce but there is no awareness to avoid such products. With use of chemical fertilizers and pesticides, natural fertility of soil is diminished and cultivators increase the doses for maintaining the productivity without understanding the future consequences.

Developed countries have banned use of chemical fertilizers and chemical pesticides in view of its adverse impact on human, animal, birds and vegetation. The soil health is also infested seriously, but developed countries have not stopped production of such fertilizers and pesticides just for running their industries and profit motive. Cultivators while spraying these chemicals are advised to cover mouth and nose with thick cloth as the smell is dangerous for health but the cultivators continue to use these harmful products for getting their crops and restrictions have so far been initiated by the government and importing these products even after understanding the consequences.

3.4 DROUGHT RECOLLECTION

Thernthwaite (1947) has classified droughts in to following four categories:

(i) Permanent Drought: which is characteristic of the driest climate, where sparse vegetation growing is adapted to drought and agriculture is possible only by irrigation during the entire crop season.

(ii) Seasonal Drought: which is found in climates which have well defined rainy and dry seasons. The natural vegetation is made up of plants that produce seeds during the rainy season and then die and plants that remain alive but become dormant in the dry season. Most of the arid and semi arid zones fall under this category. The planting dates and duration of crop varieties in those regions are so adjusted that their.

(iii) Growing season falls within rainy season followed by the period having residual moisture storage.

(iv) Contingent Drought: which is dependent upon irregularity of rainfall and so does not occur regularity in any definite season.

(v) Invisible Drought: Which can occur even when there is frequent rainfall in an area. When rains do not supply enough water to counteract the water loss by evapo-transpiration, the result is a borderline water deficiency in soil resulting in less than optimum crop yields. This occurs usually in humid area.

Barcar (1979) has broadly classified droughts in to following three categories :

(i) Meteorological Drought: It arises in a situation, when the actual rainfall is significantly less than the climatologically expected normal rainfall for an extended period of time over any large area. World Meteorological Organization (1988) has given following meteorological criterion for identification of drought situation- " 60 percent less of normal annual precipitation, for over two consecutive years for over 50 percent of the spatial extent of the region."

(ii) Hydrological Drought: Such droughts are associated with marked depletion of surface water and consequent drying of ground water resources. These water resources include precipitation, infiltration, lakes, river-systems, reservoirs, and in-flow/ out-flow systems, which may be included in the hydrological water balance equation. A hydrological drought results, if the meteorological drought is sufficiently prolonged. Hydrological drought occurs, when demand by the hydrological system greatly exceeds the water available for the use of the system. Precipitation is the most common natural factor, which usually induces drought. Precipitation has the highest spatial and temporal variability compared with other whether elements.

(iii) Agricultural Drought: It is the result of an imbalance between the soil moisture and evapo-transpiration needs of an area over a fairly long period, so as to cause damage to standing crops and a reduction in crop yield. Agricultural drought are defined using the conditions of the crop, livestock, pasture and other agricultural systems.

3.4.1 Identification of Drought Affected Areas

Drought is very vulnerable situation of disaster as the impact is widespread. This disrupts the rural economy in the area where monsoon occurrence is irregular or scanty etc. Since the monsoon crop is the main source of livelihood of the cultivator, dependence is absolute in case of the single cropped areas. In Rajasthan, about 75 percent cultivated area is rain fed, dependence on monsoon is the for the survival of the cultivators of this category. It is true that monsoon usually occurs at the specific duration say between mid-June to early-July.

Nowadays, monsoon forecast is generally communicated well in advance in view of Satellite and movement of monsoon from the sea. The forecast of monsoon arrival is known to the cultivators through media and communication measures through agriculture and irrigation department and similar organizations and the cultivators start their preparatory activities according to such forecast. Usually the forecast consist of date of expected arrival in different parts of the country and state along with the forecast of rains as normal, below normal or above normal. Cultivators are now accustomed with the intensity and usually take appropriate decision about the type of crop to be grown.

Even after the entire advance knowledge network, the information provided is common to the large area, but at local levels the spread of monsoon remains uneven and the problems of erratic nature causes many problems. Sometimes the situation remains in very scanty area, but in other conditions, the uneven situation occurs in a larger area. There are various reasons for such a situation; a few important are narrated below:

(i) Late arrival of monsoon and early departure with very few spells of rainfall occurrences.

(ii) Uneven spells along with the duration and low intensity of rains not sufficient for survival of plants.

(iii) Failure of monsoon with changes in atmospheric conditions, which could not be known in advance.

(iv) A very good start in the beginning but the following spells are disrupted for unknown reasons.

(v) Occurrence of rains with sufficient intervals and sufficient rains for survival and growth of crops but not sufficient for ground water. Such a situation hampers even the dry farm practices and problem of drinking water for human beings and livestock.

All these conditions are unusual and unforeseen. The impact of such a situation is quite significant and long duration. Since most of the cultivated area of the state is single cropped, failure of monsoon creates various problems. In such a situation, the government is required to take immediate measures to manage the situation. But before discussing all these issues, it is necessary to understand to system for identification of drought affected areas. The system prevalent in the state is narrated below:

1. Early Warning and Assessment: While there are considerable advances in space/ satellite technology for whether forecasting, but the past experiences are evident that in India and particularly in Rajasthan, the system is yet far from accuracy. Year 2002-03 proved the worst year though the long term weather forecast for 2002 monsoon by Indian Meteorology Department as normal year were completely fail to assess the situation. The short and medium term forecasting systems are yet not attained desired level of confidence. This caused by the failure of the south-west monsoon in all the 32 districts of the state, was the worst in last 136 years. Therefore traditional system of frown assessment by monitoring rainfall through the rain gause stations was further expended from Tehsil to Patwar circle to know the wide scale variations in rainfall.
2. Crop failure is assessed by revenue authorities under the instructions of Relief Department, where the District Collectors of all the districts get the information collected from Tehsildars through Patwaries posted at each Patwar circle, consisting of 1 to 10 villages on the basis of area under cultivation. Patwari collects information of crop damage for each village, where the crop damage is assessed through eye view indicating the in four categories, viz, up to 24 percent, between 25-49 percent, between 50-74 percent and 75 percent to 100 percent indicating its area. In addition, Patwari is also required to include issues like drinking water problems, fodder for livestock, etc. up to next monsoon duration.
3. This information is complied for each tehsil and for the district as a whole and the District Collector forward the information along with the recommendations keeping the crop damage in view and requisite arrangements for various activities in the district.
4. On the basis of the information received from District Collectors, State Memorandum is prepared indicating the area affected and requirement of funds for initiating the activities from the proposed date. After approval of the Memorandum from the State Cabinet, it is forwarded to Government of India for imparting assistance.
5. Generally, relief activities are carried from February to be continued up to arrival of next monsoon, but in some cases, the relief activities are carried out in earlier duration. In case of monsoon failure during 2002,

the State Government sought approval of the Government of India to start relief activities from 16-8-2002.

6. A Natural Calamity Fund is created at the state level, where the amount is decided in view of past average requirement and the Government of India provides 75 percent of its share and state contribution is 25 percent. . This fund is used to mitigate the problem. In case of severity of the problem and requirement of more funds, Government of India sends its Expert Committee to assess the situation and the quantum of assistance is decided.
7. Funding from the Government of India is received in cash and wheat for payment of wages to the people engaged in various areas. In addition, the State Government dovetails its resources for the activities mitigating the drought disaster.

3.4.2 Recollection of Droughts in Past Years

In Rajasthan, drought is a serious and vulnerable problem and its occurrence is regular in some parts or other. During last fifty years the occurrence of famine has been given in Table. 3.2 indicating the number of districts and villages affected by drought and human and livestock suffered with the problem.

Table 3.2: Extent and Intensity of Drought in Rajasthan During 1951-2003

Year	District	Village	Number of Affected		Percentage of Affected			
			Human Pop. (lacs)	Livestock Population (lacs)	District	Villages	Human Population	Live-stock Pop.
50-51	26	27300	30.57	26.78	100.00	89.00	19.14	16.21
51-52	24	27300	23.00	19.82	92.30	89.00	14.40	12.67
52-53	8	5225	13.00	9.00	30.77	14.73	8.14	6.36
53-54	11	7197	23.00	18.49	42.30	23.48	14.40	11.81
54-55	13	2162	3.71	1.99	50.00	7.58	2.32	1.02
55-56	13	1825	11.17	9.77	50.00	5.45	6.99	4.25
56.57	7	1353	0.92	0.67	26.92	4.32	0.58	0.32
57-58	14	6581	22.72	18.28	53.85	21.26	14.22	11.95
58-59	6	1309	10.77	8.63	23.07	4.14	6.74	4.86
59-60	Nil	Nil	Nil	Nil	Nil	Nil	Nil	Nil
60-61	18	4634	20.91	17.74	69.23	14.37	10.37	8.89
61-62	5	379	4.07	2.84	19.23	1.18	2.02	1.38
62-63	10	1248	7.98	5.52	38.46	3.87	3.89	2.74
63-64	18	6367	47.25	30.5	69.23	17.95	23.44	9.10
64-65	15	3156	12.85	6.52	57.69	8.89	6.37	1.99
65-66	24	11452	55.00	60.96	92.30	32.28	27.29	18.19

Year	District	Village	Number of Affected			Percentage of Affected		
			Human Pop. (lacs)	Livestock Population (lacs)	District	Villages	Human Population	Live-stock Pop.
66-67	24	10079	45.25	83.61	92.30	28.41	22.45	22.31
67-68	12	2383	15.24	21.89	46.15	6.72	7.56	5.84
68-69	29	22915	131.63	217.19	100.00	64.59	65.30	57.95
69-70	23	11977	71.72	110.76	88.46	33.76	35.28	29.55
70-71	7	480	3.02	8.53	26.92	1.35	1.17	2.28
71-72	13	6139	30.60	32.95	50.00	17.30	11.88	8.79
72-73	26	18722	136.88	184.18	100.00	52.77	53.12	47.37
73-74	Nil	Nil	Nil	Nil	Nil	Nil	Nil	Nil
74-75	25	19873	154.97	210.83	96.15	56.02	59.84	56.40
75-76	Nil	Nil	Nil	Nil	Nil	Nil	Nil	Nil
76-77	Nil	Nil	Nil	Nil	Nil	Nil	Nil	Nil
77-78	19	12253	92.45	121.99	73.01	34.54	35.88	29.49
78-79	26	5609	33.20	170.00	100.00	15.59	12.88	41.10
79-80	26	31095	240.00	301.98	100.00	86.45	93.15	74.78
80-81	26	21395	167.79	241.51	100.00	59.48	48.97	58.39
81-82	26	23426	200.12	276.12	96.30	64.63	58.41	66.76
82-83	26	22606	171.62	261.64	96.30	60.23	50.09	63.14
83-84	3	282	2.97	4.05	11.11	0.75	1.12	1.16
84-85	21	10276	92.02	132.79	77.78	27.38	26.86	26.74
85-86	26	26859	219.80	34.86	96.30	70.44	64.15	61.40
86-87	27	31936	252.70	327.49	100.00	82.54	73.75	65.96
87-88	27	36252	317.37	372.30	100.00	92.27	92.63	74.98
88-89	17	4497	43.45	83.42	62.96	11.44	12.68	20.15
89-90	25	14024	120.67	132.29	92.59	35.35	35.22	32.34
90-91	Nil	Nil	Nil	Nil	Nil	Nil	Nil	Nil
91-92	30	30041	289.00	270.93	100.00	74.60	74.41	66.24
92-93	12	4376	34.66	42.46	40.00	10.87	8.92	8.89
93-94	25	22586	246.81	259.64	83.33	56.00	56.08	54.35
94.95	Nil	Nil	Nil	Nil	Nil	Nil	Nil	Nil
95-96	29	25478	273.82	282.29	93.55	62.47	62.22	59.09
96-97	21	5905	55.29	72.60	67.74	14.37	12.56	15.20
97-98	24	4633	36.28	45.91	75.00	11.27	9.02	9.05
98-99	20	20069	215.07	295.78	62.50	48.83	48.87	54.42

Year	District	Village	Number of Affected			Percentage of Affected		
			Human Pop. (lacs)	Livestock Population (lacs)	District	Villages	Human Population	Livestock Pop.
99-00	26	24406	261.79	345.61	81.25	59.38	59.49	63.59
00-01	31	30583	330.41	399.69	96.87	75.78	74.75	78.63
01-02	18	7964	69.70	69.73	56.25	19.74	16.49	14.58
02-03	32	40990	447.80	451.63	100.00	96.98	98.78	93.94

Source: Department of Disaster Management and Relief, Govt. of Rajasthan.

It is evident from the above information that during the period of last 53 years, there was no famine in any part of the state during six years, i.e., 1959-60, 1973-74, 1975-76, 1976-77, 1990-91 and 1994-95. Similarly during 10 years all the districts were declared famine affected. More than 75 to 99 percent districts were affected by famine during 14 years. 50 to 74 percent districts were affected by famine during 12 years. The intensity of drought was lesser during 13 years as less than 50 percent districts were affected by famine conditions.

The intensity of drought has been assessed by number of persons affected by famine conditions and for this purpose the drought conditions have been sub-divided in five categories. According to information available with the State Government for past 103 years it is evident that during 16 years the drought conditions were very serious in 16 years, severe drought was witnessed during 9 years, moderate drought was faced in 16 years, light drought occurred in 18 years and during 44 years, the drought conditions were normal in view of its intensity and spread. This information has been given in Table 3.3 below:

Table 3.3: Intensity of Droughts in Rajasthan (1901-2003)

Intensity	Years	Total No. of Years
Very Severe	1901, 1905, 1911, 1915, 1918, 1951, 1952, 1969, 1980, 1986, 1987, 1988, 1992, 1996, 2001, 2003	16
Severe	1902, 1904, 1913, 1921, 1925, 1939, 1975, 1982, 2000	9
Moderate	1907, 1920, 1936, 1938, 1941, 1953, 1966, 1967, 1970, 1973, 1978, 1981, 1983, 1990, 1994, 1999	16
Light	1922, 1929, 1930, 1932, 1940, 1948, 1949, 1954, 1958, 1964, 1972, 1979, 1985, 1989, 1993, 1997, 1998, 2002	18
Normal	1903, 1906, 1908, 1909, 1910, 1912, 1914, 1916, 1917, 1919, 1923, 1924, 1926, 1927, 1928, 1931, 1933, 1934, 1935, 1937, 1942, 1943, 1944, 1945, 1946, 1947, 1950, 1955, 1956, 1957, 1959, 1960, 1961, 1962, 1963, 1965, 1968, 1971, 1974, 1976, 1977, 1984, 1991, 1995	44
	Total	**103**

Source: Department of Disaster Management and Relief, Govt. of Rajasthan.

Drought of Samvat 2059 (year 2002-03) was the most severe drought of recent year and the extent of severity has been adjudged on the following indicators, giving district-wise information as per Table 3.5.

Table 3.5: Extent of Scarcity of Drought- *Samvat* 2059

District	Total Villages	Affected villages			Pop. affected	Area Sown
		50-74 %	75-100 %	Total	(Lacs)	(lac. Ha.)
1	2	3	4	5	6	7
1. Ajmer	1076	11	1065	1076	11.78	3.06
2. Alwar	1984	123	1857	1980	27.65	2.45
3. Banswara	1524	1510	—	1510	13.49	2.23
4. Baran	1233	688	519	1207	13.76	1.19
5. Barmer	1975	1	1974	1975	14.35	9.77
6. Bharatpur	1505	—	1504	1504	1.83	1.70
7. Bhilwara	1788	290	1494	1784	14.71	3.20
8. Binaker	889	—	889	889	19.01	1.24
9. Bundi	870	369	501	870	8.86	1.60
10. Chittorgarh	2415	1193	1151	2344	16.34	3.89
11. Churu	907	12	895	907	13.73	4.32
12. Dausa	1061	117	939	1056	11.93	1.44
13. Dholpur	828	3	825	828	8.87	0.71
14. Dungarpur	872	—	872	872	8.74	1.09
15. Ganganagar	3028	484	1668	2152	7.05	1.53
16. Hanumangarh	1912	314	1525	1839	11.50	3.29
17. Jaipur	2340	35	2237	2272	26.16	4.89
18. Jaisalmer	688	36	607	643	4.43	0.24
19. Jalore	728	5	723	728	11.43	4.44
20. Jhalawar	1614	325	1266	1591	10.70	2.68
21. Jhunjhunu	867	—	867	867	16.00	2.32
22. Jodhpur	1082	36	1046	1082	24.06	1.02
23. Karauli	829	9	820	829	9.73	1.31
24. Kota	947	521	379	900	9.05	1.28
25. Nagaur	1508	34	1474	1508	19.49	5.87
26. Pali	973	8	965	973	14.76	3.23
27. Rajsamand	1004	10	994	1004	8.03	0.86
28. S.Madhopur	800	62	732	800	11.16	1.05
29. Sikar	1017	6	1.11	1017	18.45	3.80

District	Total Villages	Affected villages			Pop. affected	Area Sown
		50-74 %	75-100 %	Total	(Lacs)	(lac. Ha.)
1	2	3	4	5	6	7
30. Sirohi	474	—	474	474	6.58	1.16
31. Tonk	1122	49	1054	1103	10.99	3.02
32. Udaipur	24.06	957	1449	2406	25.77	2.25
Total	**42266**	**7208**	**33782**	**40990**	**447.80**	**82.26**

Table continued ...

District	Affected Area (Lakh Ha.)	Affected Cattle (Lacs)	Value of Damaged Crop (crore Rs.)	Land Revenue Suspended (Lac Rs.)	Rainfall Deficit %
1	8	9	10	11	12
1. Ajmer	2.97	7.67	183.62	16.29	65
2. Alwar	2.16	15.41	129.36	33.46	72
3. Banswara	1.30	11.82	76.66	8.99	38
4. Baran	0.91	6.73	86.82	16.88	62
5. Barmer	9.67	42.02	220.09	—	69
6. Bharatpur	1.70	7.96	84.85	21.97	50
7. Bhilwara	3.03	19.77	203.46	38.27	56
8. Binaker	1.01	23.17	10.07	13.10	85
9. Bundi	1.23	9.62	87.51	10.27	56
10. Chittorgarh	2.80	17.16	203.56	21.65	51
11. Churu	4.24	18.10	121.63	—	64
12. Dausa	1.28	6.56	58.63	8.88	62
13. Dholpur	0.59	4.53	43.00	5.32	44
14. Dungarpur	0.93	9.58	9.32	0.87	37
15. Ganganagar	1.13	8.80	1.77	27.42	76
16. Hanumangarh	2.79	8.98	—	—	79
17. Jaipur	4.37	20.36	274.66	51.26	71
18. Jaisalmer	0.22	12.56	4.34	2.59	83
19. Jalore	4.30	6.26	260.47	43.88	59
20. Jhalawar	1.47	8.62	113.91	8.50	46
21. Jhunjhunu	2.27	12.43	75.85	4.03	75
22. Jodhpur	—	33.07	279.86	—	76
23. Karauli	1.23	6.38	81.74	16.62	69

District	Affected Area (Lakh Ha.)	Affected Cattle (Lacs)	Value of Damaged Crop (crore Rs.)	Land Revenue Suspended (Lac Rs.)	Rainfall Deficit %
1	8	9	10	11	12
24. Kota	0.88	3.73	139.52	35.05	53
25. Nagaur	5.40	19.57	380.65	—	69
26. Pali	3.16	28.04	301.56	—	68
27. Rajsamand	0.76	11.93	81.11	10.32	49
28. S. Madhopur	0.93	8.18	50.75	—	71
29. Sikar	3.60	16.63	197.99	9.87	67
30. Sirohi	1.11	9.83	72.10	—	71
31. Tonk	2.86	9.88	118.75	8.26	70
32. Udaipur	1.67	26.28	77.66	15.49	36
Total	**71.97**	**451.63**	**4031.37**	**429.26**	**64**

Source: Department of Disaster Management and Relief, Govt. of Rajasthan.

Information given above reveals that the intensity of drought was utmost severe as all the districts were affected with late arrival and later failure of monsoon as 40990 villages out of 42266 were affected. In 7208 villages the crop damage was between 50 to 74 percent and in 33782 village it was 75 to 100 percent. Out of the crop area sown of 82.26 lac hectares, 71.97-hectare crop area was damaged. The value of crop damage was estimated to be Rs. 4031.37 crores (Fig. 24, 25 and 26).

These all aspects need to be spelt out in the memorandum to be submitted to the Government of India for getting assistance to combat the drought disaster. The quantum of assistance required from the Government was indicated as Rs. 6115 crores, which contained employment generation activities as Rs. 4600 crores – Wage component – Rs. 1036 crores, wheat 56 lacs MT amount Rs. 2324 crores and material component for relief projects Rs. 2240 crores. In addition other requirements were -cattle conservation – Rs. 165 crores, drinking water supply Rs. 300 crores and medicines Rs. 50 crores. This amount was subsequently revised to Rs. 7519 crores at the time of visit of Central Study Team. The earlier proposed requirement of funds was completely revised in view of the gravity of the relief as under:

1. Employment Generation of 56 crore mandays which included food grain requirement of 56 lacs MT- Rs. 5348 crores
2. Under cattle conservation for Transport subsidy for fodder of 24.80 lacs MT - Rs. 379.25 crores
3. Assistance to goshala for cattle conservation and organization of cattle camps - Res.545.12 crores.
4. Bull subsidy for maintenance of high breed 8834 bulls- Rs. 3.18 crores.

5. Cattle feed subsidy to 25 lac cattle- Rs. 271.50 crores.
6. Drinking water requirement- Rs. 518.05 crores.
7. Medicines at relief sites- Rs. 10.00 crores.
8. Input subsidy to small and marginal farmers number 23.46 lacs- Rs. 234.56 crores.

This is the total process, which is being adopted by the state government when the drought conditions prevail and the state government is not in a position to manage the work from its own funds and through the amount available.

4

Drought-Extent, Causes, Processes and Indicators of Measurement

Drought is one of the most critical disasters, which occurs due to failure of monsoon in many ways. Categories of monsoon failure are varied starting from late arrival, disturbances in regular rainfall spell and early departure. From drinking water and charging sub-surface water resources, occurrence of rainfall in total monsoon duration has relevance but from cultivation point of view, timely arrival, frequent rainfall spells and timely departure are critical from germination and growth of crop. Agriculture is quite sensitive issue and cultivation of rain-fed crops is further critical as there are various factors having positive and negative impact. From cultivation point of view, low showers in each spell are sufficient for crop production with low productivity but if all the rainfall spells are timely, cultivation is possible.

Even irrigated areas receiving water in the fields in time could not yield similar productivity, which is possible with rainfall, which changes total environment conditions of the region. Productivity is increased to sizeable extent as rains not only supply water to plants in need but also changes total environmental scenario favourable for cultivation needs. However, nature plays its own role as per geographical and environment factors and accordingly monsoon has close relationship with agriculture practices. Atmospheric changes due to rainfall have positive impact over the cultivated crops. Thus relevance of monsoon in kharif crops is realistic and crop productivity is maximum when occurrence of rainfall is perfect to agricultural needs.

This is also relevant to consider that occurrence of rain is varied in various regions of Rajasthan. Normal annual rainfall in Jaisalmer and Banswara is 18.55 cm and 95.03 cm respectively. If Jaisalmer district can coup up in very little rainfall, Banswara district can also manage its water requirements in same rains. This is particularly associated with geographical and meteorological features and both the districts are sustaining in their own way as people of both the regions and all the regions have become accustomed with such environmental conditions. Deviation from the normal conditions is disastrous for all the regions and geo-physical conditions have become favourable to these phenomena.

4.1 AGRO-CLIMATIC ZONES

Before analyzing the drought conditions in the state, it is relevant to discuss about various climatic zone of the state as climatic conditions of a region affect the agricultural cropping pattern and various areas grow different types of crops best suited to environmental conditions. Amongst a host of climatic factors, rainfall, temperature, humidity, wind velocity and duration of sunshine etc affect the cropping pattern in a significant manner. Annual rainfall, its distribution over the entire year, regimes of diurnal and annual temperatures are the prominent factors affecting the agriculture and the life style of the people. On the basis of climatic conditions and agricultural produce, Rajasthan has been divided into nine agro-climatic zones, each one having special characteristics of its own.

4.1.1 Arid Western plains

This region comprises of all tehsils of Bikaner, Jaisalmer and Barmer districts, Phalodi, Shergarh, Osian and Jodhpur tehsils of Jodhpur district, Dungargarh, Sujangarh, Ratangarh and Sardarshahar tehsils of Churu district. This is most arid part of the state where annual rainfall varies from 10 to 40 cm, quite often erratic, so much so, the entire rainfall of the year may fall on a single day and rest of the year may be dry. Summer temperatures are always high and the diurnal range exceeds even 20° C. During the day, the summer temperatures may be as high as 49° C but in the night, the temperatures may fall to less than 20° C. In winters, the day temperatures are higher but the night temperatures may be near freezing point. Winters are of short duration not exceeding two months- December and January.

This is an area of not so developed drainage system where there are no flowing streams. Owing to poor rainfall, surface water resources do not exist while ground water resources are often deep and brackish. Normal vegetation is therefore, only seasonal. With the first showers in late July, few grass species grow, shrubs and dwarf trees become green. But soon after the retreat of monsoon in mid September, the vegetation dries up, leaving only few perennial shrubs and a thin pad of pale grasses on landscape. Most rain-fed crops are Bajra, kharif pulses, guar etc are grown during kharif season. Rabi crops like wheat, rape seed and mustard are possible in areas where irrigation facilities are available.

4.1.2 Irrigated North-Western Plains

Entire Ganganagar and Hanumangarh districts, which are diurnal and Aeolian plain, formed by the river Ghaggar, the ancient river Saraswati, forms this agro-climatic zone. The Ghaggar channel even though 7 kms wide at some places, inundates nearby villages during heavy rainfall in its catchment areas. A part of this region which is arid, is the northern extension of the Indian Thar Desert covred with wind blown sand. Extreme aridity marked with high summer and low winter temperatures, is its usual climatic characteristics. The average annual rainfall is about 40 cms. High speed winds blow in summer. The area is rich in agricultural produce on account of well-developed canal irrigation system. A large network of Gang Canal, Bhakra canal and Indira Gandhi Canal has made the entire area green and productive.

Due to abundance of canal water irrigation, this region has become the greenery of Rajasthan. Amongst the kharif crops cotton, sugarcane and pulses are main crops being sown. In rabi season wheat, mustard, gram, vegetables and fruits are produced. Total production and productivity of all crops are relatively much higher in this zone as compared to other parts of the state. For last three decades, this area is seriously affected with water-logging problem, which has put adverse impact on the cropping pattern. Cultivators in vicinity of canal irrigation used abundant water for crops and now the problem has developed to great extent.

4.1.3 Transitional Plains of Inland Drainage

This zone comprises all tehsils of Nagaur, Sikar and Jhunjhunu districts, Taranagar, Churu and Rajgarh tehsils of Churu district. This area is covered with sand dunes and inter-dunal sandy plains. Drainage system is not well developed and streams which flow in the rainy season, disappear in sandy fields after covering some distance. Climatically, this zone is slightly better as compared to the adjoining zone of the arid western plains. Rainfall is slightly higher, temperatures in summer months reach to very high level but the winters are very cold. Irrigation is restricted to areas with good ground water potential. Bajra, sesamum and kharif pulses are the main crops of rainy season. Wheat, barley, mustard and gram are grown as irrigated crops or on conserved soil moisture during rabi season.

4.1.4 Transitional Plains of Luni Basin

This area lies between the Aravalli ranges and western arid regions. The region comprises of the entire districts of Jalore and Pali, Reodhar and Sheoganj tehsils of Sirohi district and Bilara and Bhopalgarh tehsils of Jodhpur district. This region has semi-arid climate with annual rainfall of 30 to 50 cms. It is drained by river Luni which is seasonal and flows during rainy season. A number of paleo-channels also exist in this area. Western part of this region is dotted with sand dunes, interspersed in alluvial soil. Luni and its several tributaries like Sukri, Mithri and Jawai have made this area productive. The climatic conditions are almost the same as in the western arid region except that the rainfall is slightly higher. The ground water level is high in the river basins and has been usefully tapped for irrigation. Vegetation is xerophytic and sparse in the western part but in the east and on the slopes of the Aravalli ranges, there is mesophytic vegetation in the form of woodlands, open forests and grasslands. The area produces bajra, maize, guar, sasamum and pulses in the kharif season. In the rabi season, wheat, barley and mustard are the dominant crops specially in irrigated areas.

4.1.5 Semi-arid Eastern Plains

This region comprises of Jaipur, Dausa, Tonk and Ajmer districts. Banas with its several tributaries form a rich fertile plain. On the western side, the region is flanked by the low Aravalli hills, which extend from the south-west to north-east. Annual rainfall of the region varies from 50 to 60 cms with an increasing trend towards east. Summer and winter temperatures are not as extreme as in the arid west but summer temperature usually reaches around 45° C and in winter, minimum temperature falls up to 8° C. The water table varies from 15 to 25 metres but annual fluctuations are high specially in the years when south-west monsoon fails and the yearly replenishments are low. Surface water sources are scarce and so harnessing of ground water resources have gone to an accelerated pace. Natural vegetation is of the mixed xerophytic and mesophytic type, but owing to heavy felling of trees, the surface mantle has been robbed of its natural wealth. In the total gross cultivated are of this zone, bajra, sorghum and pulses are grown in the kharif season and wheat, barley, gram, mustard in the rabi season. Productivity of all crops in this zone is better than that of the agro-climatic zones that are to the west of the Aravalli ranges.

4.1.6 Flood Prone Eastern Plains

This region comprises of entire districts of Alwar, Bharatpur, Karauli and Dholpur and Gangapur, Bonli and Bamanwas tehsils of Sawai Madhopur district. The entire region is flood plain of Banganga and Gambhiri rivers except few low hills existing in Alwar, Karauli and Sawai Madhopur districts. The region has rich alluvial soils the fertility of which is replenished every year by the flood water of the seasonal rivers. In periods of heavy rainfall, the rivers outflow their banks and inundate the surrounding villages.

Climatically, the area is similar to the plains of Banas but the rainfall is relatively higher in the east, the annual average rainfall being about 75 cms. Natural vegetation exists on mountain slopes, wetland areas and protected zones but the excessive plundering of forest wealth has dwindled the natural cover. The region produces variety of crops as ground water and surface irrigation sources exist. A network of canals drawn from upper Yamuna Canal and the Panchana dam irrigate this area. The ground water aquifers vary from 5 to 15 metres which helps in well irrigation practices. The region produces bajra, sorghum, maize, sugarcane, sesamum and a variety of pulses in kharif season. Wheat, barley, gram and mustard are dominant crops of rabi season.

4.1.7 Sub Humid Southern Plains and the Aravalli Hills

Bhilwara and Rajsamand districts, all tehsils of Udaipur district except Dhariyawad, Salumbar and Sarada tehsils, all tehsils of Chittorgarh district except Chhoti Sadri, Pratapgarh, Arnod and Bari Sadri tehsils, Abu Road and Pindwara tehsils of Sirohi district form this agro-climatic zone. For the most part, this is a region of low Aravalli hills with the inter-mountain plateau, deeply dissected by streams and rivers. The region has a moderate warm climate in summer and mild winers. Annual rainfall varies from 50 to 95 cms. Highest precipitation in the state is recorded in Abu hills of Sirohi district. There are number of surface water streams like Ghambiri, Sabarmati, Banas and its tributaries but all are ephemeral.

The surface rocks are granitic and highly metamorphosed where depressions on he surface are filled with rain water in the form of natural tanks. The area is rich in natural vegetation, which grows on the slopes of the Aravalli hills and wetland areas but excessive felling of trees has degraded these open forests. Tank water irrigation is most common. The area produces maize as chief food crop of kharif season but paddy growing is also possible in irrigated areas. In rabi season wheat, gram and oilseeds are main crops. In areas of black soil, cotton and opium cultivation is also carried out.

4.1.8 Humid Southern Plains

Districts of Dungarpur and Banswara, Dhariawad, Salumber and Sarada tehsils of Udaipur district, Chhoti Sadri, Bari Sadri, Pratapgarh and Arnod tehsils of Chittorgarh district are included in this region. The area comprises of low Aravalli hills with intrusion of black lava rocks. This is mostly tribal area where Bhils, Garasiyas and Damors live amidst forests and hills. The area has humid climate with an average rainfall of more than 70 cm per year. The temperature regimes do not fluctuate much in summer and winters and harshness of temperature does not prevail. Humidity is always high and all factors combined together promote a profuse growth of natural vegetation. Mountain slopes have an abundance of natural forests but uncontrolled felling of trees by outsiders is degrading the environment of this vulnerable region.

There are innumerous surface water streams. Mahi and its tributaries like Anas, arau and Jhakham have made this area very fertile, except few exceptions. Commissioning of Mahi Bajaj Sagar multipurpose project has provided this area with canal irrigation and electric resources. Cotton and sugarcane are chief cash crops grown in black soil region, which is significantly fertile but in case of low rain-water, the soil structure remains very hard to plough. Maize, sorghum ad sugarcane are main food crops of kharif season. Ground nut, mustard, sesamum and rapeseed are also grown.

4.1.9 Humid Southern- Easter Plains

Popularly known as Hadoti plateau, this region includes districts of Kota, Baran, Bundi and Jhalawar, Khandar and Sawai Madhopur tehsils of Sawai Madhopur district. Being rocky area there is very thin layer of black soil over hard soil structure, which cultivators try to retain for cultivation purposes. However the black soil plateau region is fertile and used for cultivation of sugarcane, cotton and opium. The region has low hills of Gwalior series, interspread with broad plateau of Vindhyan rocks. Large number of rivers drain in this region where Chambal is main perennial river and its main tributaries are Parvati, Kali Sindh, Parwan and Banas but all are seasonal rivers disappear after rains. Development of canal irrigation system with series of dams and barrages on Chambal made this area richin agricultural production. In the present time many medium irrigation projects and anicuts have been constructed outside state over Chambal river created serious problem of availability of requisite water. Gandhi Sagar, Rana Pratap Sagar and Jawahar Sagar dams together with Kota Barrage are hydro energy sources of state and canal systems for irrigation. These dams are hardly filled in to total capacity in normal rainfall conditions due to low discharge of water from neighbouring state.

The region has warm summers but mild winters. Summer temperature usually touch 45° C but relative humidity is generally high in this zone. Annual rainfall varies from 60 to 85 cms. Natural vegetation exists in the form of woodlands, parklands and open forests, but now degraded to serious extent by uncontrolled felling. Paddy and sorghum are chief food crops of kharif season. This area is suitable for soyabean crop also. Wheat, barley, gram and mustard are grown in winter season.

4.2 EXTENT OF THE PHENOMENON OF DROUGHT

Drought is most disastrous calamity specially in hyper-arid, arid and semi-arid regions of the State and total state is prone to drought. This phenomenon is directly associated with rainfall which is completely erratic and uncertain. Chances of failure of monsoon are vital in the state and total state is prone to drought conditions irrespective of its extent and intensity. There is every possibility of drought conditions in one part or the other depending on the occurrence of rainfall. The rainfall remained most erratic and failure of monsoon is regular feature in the state. Before occurrence of rainfall in the state in monsoon season, people of the state remained feared from period of arrival in June and up to departure of monsoon in the month of September. The cultivation practices remain at the stake for total duration as behaviour of monsoon remains completely uncertain. Cultivators are main sufferers of monsoon failure and half of the cultivator families have very scanty land resources for cultivation and their livelihood remains uncertain till the harvested crop reaches to home.

Rajasthan receives precipitation from two systems where the south-west monsoon systems develop over Indian Ocean in middle of May and touches Kerala coast around first week of June every year. Further forecasts of arrival of monsoon are estimated in terms of probable date. There remains total uncertainty when the monsoon advances further towards Indian sub-continent in to two branches where one moves through Bay of Bengal and another through the Arabian Seas. The whole monsoon system is governed by different atmospheric and meteorological conditions as well as formation of low-pressure zones over sea and land surface.

The onset of monsoon is directly related with the atmospheric activity governed by external forces unable to be predicted in advance. These forces include solar and terrestrial radiation, mountains and influence of oceans, which help in creation of monsoon. Its further migration from one place to other depends upon differential heating of the region, which creates low-pressure zone and invite air loaded with high proportion of humidity from areas of high pressure.

Monsoon in Rajasthan state is governed by various factors inherent to state peculiarities. Aravalli mountain ranges are main obstacle for monsoon in western region. In addition dust storms in summer months rise high create further obstacle in rainfall in desert region. Usually low pressure do not activate to attract clouds which help in diversion of thick clouds to other regions. Usually the meteorologists believe that monsoon occurrence is full in total country but its dispersal remains varied through out the country. Rajasthan state is deficient in rainfall while some regions of the country are prone to floods creating serious havoc of devastation. There are innumerous geographical and meteorological factors responsible for deficient rainfall.

Extent of drought is most common as one or other part is facing drought situation. There are very few exceptional years when total state remains free from the drought conditions. Soil structure of the state is quite varied and heavy downpour creates severe devastation in desert region where drainage system is mostly absent. Situation of Barmer and Bikaner districts is well known to the country where rainwater created severe flood position in 2006 rainfall and stored water could not be cleared for more than 8 months after heavy downpour.

The extent of drought remains varied from a part of any district of the state to serious drought situation in almost all the districts covering largest of the state where few exceptional villages are left with the problem. Occurrence of rainfall on yearly basis has no relevance as drought conditions are possible with only 3 percent rainfall deficit. The actual problem of Rajasthan state is timely arrival of monsoon, its regular spells and timely retreat. If these spells are maintained, even drought conditions do not prevail in the state. In the year of 1999 there was only 3 percent deficit rainfall and 20069 villages of 20 districts out of total 41353 villages of state were affected with serious drought conditions. Total 215 lakh human population and 295 livestock population was affected with 64.96 lakh hectare cropped area and total estimated crop damage was worth 2283 crores.

Similarly in the year 1999-2000 23406 villages of 26 districts were drought affected with 16 percent rainfall deficit affecting 262 lakh human and 345 lakhs livestock population. In this year crop 78.18 lakh hectare area crop was damaged valuing Rs. 3407 crores. In the year 2000-01 30583 villages of 31 districts were drought affected with 29 percent rainfall deficit where 330 lakh human and 400 lakh livestock population was affected and crop worth Rs. 3512 crores was damaged in 89.47 lakh hectare of area In 2001-02 there was 5 percent rainfall deficit causing drought conditions in 7964 villages of 18 districts affecting 70 lakh each of human and livestock population. Crop standing in 26.53 lakh hectare area worth Rs. 1252 crores was damaged. Most serious drought occurred in 2002-03 affecting 41000 villages of 32 districts where 432 lakh human and 543 lakh livestock population was affected. Crop worth Rs. 4414 crores was damaged in 117 lakh hectare area.The rainfall deficit during the was 64 percent.

During the year 2005-06 drought affected 15778 villages of 22 districts affecting 198 lakh human and 178 lakh livestock population. Crop of 55.03 lakh hectares area was damaged valued worth Rs. 2562 crores. The rainfall deficit of state as whole was 5.42 percent while varied from less than one percent to 55 percent. This is quite surprising that in Nagaur district the rainfall deficit was only 0.17 percent but crop worth Rs. 202 crores was damaged. Thus occurrence of rainfall has little relevance but most critical component of rainfall is timely occurrence which is the ultimate factor of drought conditions. In such situation, drinking water problem is not serious but crop damage and people affected is problematic.

During the year 2006-07 state rainfall was 15 percent over normal rainfall but affected districts had deficit ranging between 3 to 45 percent. This year 10529 villages of 22 districts were affected with drought where 42 lakh hectare area crop was damaged valued Rs. 3992 crores. 137 lakh human and 142 lakh livestock population was affected from drought. For calculation of drought affected cropped area and amount less than 50 percent damage is not accounted for as it is treated as normal phenomenon. If total damage of crop and populated affected is calculated, that can be at least three times more in affected areas. The norms set by the Government of India refer to damage ranging between 50 to 74 percent as partial damage and 74-100 percent as full damage.

Rainfall and drought are positively correlated but more relevance is the occurrence of rainfall at the time of requirement. Usually crops are sown after first rains and cultivators wait till arrival of monsoon. In case of late arrival, cultivators either left their cultivated fields as fallow or try to sow crops species possible in short duration. Early retreat is unable to be guessed as that depends on the atmospheric conditions remain changing due to various climatic and wind factors. In case of early withdrawal of monsoon cultivators are able to collect fodder from their fields to feed their livestock as possible measures.

Rajasthan state has large area with varied soil structure, temperature, climatic conditions and occurrence in the state is significantly varied. There is wide variation in the quantum of rainfall and each region has adjusted its lifestyle as per varied factors. Normal rainfall in Jaisalmer and Banswara districts is largely varied but cultivation practices have been adjusted by the people accordingly. Even Jaisalmer district with extreme temperatures is able to grow crops through monsoon rains and also through moisture retention with very low rainfall. There is more than four times variation in normal annual rainfall, but occurrence of drought has equal possibilities in all the regions irrespective of quantum of rainfall. Every agro-climatic zone has managed its agriculture cycle according to occurrence of rains in their respective regions.

Analysis of past years reveal that rainfall has relevance with drought but even low rainfall of timely accruals can help in receiving good crops and even minor variation in normal annual can create drought conditions if rains are not in accordance with the agriculture requirements. With erratic and uncertain behaviour of monsoon, cultivation practices highly suffer and most critical factor is timely and regular spells of rains. The people of Rajasthan state are bravely facing all the adversities and do not try to migrate for other areas of the country as cultivators are helpless with their conditions and not able to chose alternative sources of livelihood. Presence of large number of livestock is the survival base of poor agriculture practitioners who try to establish then by this assured income source and helpful to face droughts. However the pressure on land is increasing tremendously to manage it.

Extent of drought in the state is quite severe being a creeping hazard which develops slowly but prolong sometimes over many year. The affected area is able to coup up in more time as the consequential impact take time to recover. Loss of moisture decreases the moisture contents from sub-surface soil helpful in germination and growth of plants. The recovery period may be quite large depending on the intensity and impact. Similarly loss of vegetation cover takes long time even if sincere efforts for replantation and revegetation are carried out. Usually environment and ecology loss is not of any person and revival process is difficult. Most of the land surface of the earth in the state has reason to no-recovery as people do-not deem proper to regenerate the common property or community resources though ultimate beneficiary are the people of that specific habitation.

The extent of drought is known from the damage of crops and scarcity of water in various parts of the state. In case of crop damage less than 50 percent in any region is treated as normal condition as damage up to this extent is deemed to be borne by the cultivators and measures for supporting the people are taken. If the crop damage is more than 50 percent and up to 74 percent, it is treated as severe drought and supportive measures are taken as this condition is treated as unbearable to cultivators and other affected people. Crop damage more than 75 percent to 100 percent is treated as most severe drought and various supportive measures are initiated. The frequency is also relevant as in case of affect in very smaller area, the supportive measures are restricted to that particular region or regions but when more than half to complete state is severely affected with drought conditions, total administration is alarmed to take various measures as per requirement.

Occurrence of most severe drought has also been analyzed on the basis of past frequency of such droughts and it has been estimated that five districts of the state viz. Barmer, Jaisalmer, Jalore, Jodhpur and Sirohi have three years of frequency of most severe drought. Occurrence of most severe drought in 8 districts viz. Ajmer, Bikaner, Bundi, Dungarpur, Ganganagar, Hanumangarh Nagaur and Churu is four years. Ten districts of the state- Alwar, Banswara, Bhilwara, Jaipur, Jhunjhunu, Pali, Sawai Madhopur, Sikar, Dausa and Karauli face most severe drought conditions once in five years. Occurrence of most severe drought once in six years is expected to be faced by seven districts of the state- Chittorgarh, Jhalawar, Kota, Udaipur, Tonk, Rajsamand and Baran. Bharatpur and Dholpur districts of the state are facing most severe drought conditions once in eight years. The analysis of occurrence of most severe drought in past 54 years from 19550-51 to 2002-03 reveal that 21 years were most critical for the state where almost total state was completely gripped by drought.

4.3 CAUSES AND PROCESSES OF DROUGHTS

Drought is a creeping hazard as it develops slowly due to failure of monsoon but has prolonged existence sometimes for many years. Drought is not constrained to a particular tectonic topographic setting but their impacts are extended over very larger part of the state. Drought cannot be measured by low occurrence of rains in total duration of the year but has its multi-faceted factors affecting the area. The crop failure is one of the process which hits the cultivators to great extent. Drought has most serious impact when it is converted into famine, which destroys food security to the point where large number of people struggle for food for survival, resulting mass deaths from starvation. However, the relationship between drought and famine is not always direct as drought is a geophysical hazard whereas famine is a cultural phenomenon, which are sometimes resultant of crop failure at large extent.

Drought is treated as geophysical factor where its occurrence is the resultant impact of failure of monsoon. Usually monsoon failure is treated as the ultimate cause of the drought and supportive measures are initiated the factors which are affected most. In case of crop damage, wage employment measures are initiated. In case of drinking water crisis, arrangements are made to provide water to problematic areas. In case of starvation, food supply and medical support is managed to save people from mass deaths. All such measures are precautionary measures to provide relief to affected population. In case of fodder shortage, arrangements are made to procure fodder from surplus states or regions and fodder depots are established. Cattle are also preserved in goshalas to protect from starvation. These are social factors which are adopted to provide suitable relief in the affected areas and all possible arrangements are made by the government treating it as social responsibility.

The process of drought starts from failure of monsoon treating it a natural phenomenon. There is no doubt about the occurrence of monsoon in various regions is due to various natural factors, but creation of adverse conditions for monsoon failure is also associated with human interface with the natural factors, which is most important to analyze to assess the conditions which are responsible for failure of monsoon. There are various factors of human interface, which have deteriorated the environment and ecology of the state and monsoon failure is the resultant impact of these factors also. Analysis of past monsoon occurrence with the present level of deterioration reveals that human acts have created the conditions which are also responsible for failure of monsoon or uncertainties.

4.3.1 Uncontrolled Deforestation

Large scale deforestation is the prime cause of drought as monsoon is attracted with the vegetation. It is very relevant to mention that southern Rajasthan had very favourable monsoon and lakes were full during the rainy season. Now the most of the mountains are denuded causing uncertainty of monsoon in the region. Similarly western Rajasthan was totally green area some 5000 years back when River

Saraswati was flowing through the desert region. There were thick forests in total area, but now total area is covered with sand dunes, denuded mountain ranges and low hills. Uncontrolled felling of trees has created large scale devastation. Forests help in maintenance of the ecosystem of any area by keeping the surface soil structure intact, moisture conservation, regulating the temperature, decreasing the evaporation process but by felling of trees from the total regions, the adverse impacts started creating problem in the form or erratic and uncertain monsoon.

People require fuel wood, fodder and timber for various reasons. Fuel wood is the necessity of the people, which can be met through agriculture residue, cutting dead trees, shrubs and bushes, but cutting of live trees developed as a process for meeting all human requirements and control measures for protection of trees could not prevent such losses at large scale. Every destroyer treated to fell one tree from large and thick forest to keep its act ineffective but the situation remained uncontrolled as people never thought about the long term consequences of their destructive acts. Grazing of livestock in open and cutting branches of trees initiated the process which also continued and intensified during the droughts. Similarly requirement of wood for industrial uses remained part of development process and continued with administrative and political connivances. Therefore, without pinpointing the destructive human acts, the forests and plantation areas have been vanished.

Soil erosion in mountain regions and wind storms started creating havocs as there remained no barriers to heck the natural factors. The decrease of rainfall is not a process of one year or one decade but consequential impact of continuous destructive measures of people. Evaporation process remained uncontrolled causing disappearance of water bodies and earth surface started deteriorating losing its productivity. People also realize that deforestation has created adverse impact over the cultivation process as soil erosion, movement of dunes through wind storms lost the drainage systems and loss in cultivated area. About half of the cultivable land has become uncultivable causing serious loss to cultivating community, which they realize as consequential impact. Eastern Rajasthan became flood prone and productivity of cultivated land has been seriously affected.

4.3.2 Uncontrolled Grazing

During rainy season, grass is grown over sand dunes, inter-dunal plains, mountains, wasteland and barren lands. Grass seeds are dispersed through wind and germination process start from seed with occurrence of rains. In desert regions, total surface area is full of vegetation, which is vanished by grazing livestock, which the livestock raisers use without knowing consequences of their acts. The vegetative cover helps in keeping the soil structure intact from winds but with uncontrolled grazing, the upper surface soil got cracked allowing wind to play its role. Sheep, goat, camel and cattle are main source of enmass destruction and local community remained spectator allowing such acts as the loss remained over the common or community land resources. Sheep, goat and camel are migratory livestock creating serious problem to vegetative cover.

4.3.3 Mechanization Process

Mechanization process in agriculture and movement of vehicles created serious problem in disrupting the soil strata. Movement of tractors, trucks, buses on hard soil cover crippled the earth surface allowing wind to disrupt the fertile soil movement from cultivated fields and surrounding areas. Ploughing through tractors is dangerous but mechanization is also essential for quick cultivation process. With constant movement of mechanized vehicles, moisture of soil is also disturbed to significant extent. This process helped in early movement and application of various practices, which took long time with human or livestock support.

4.3.4 Application of Technological Devices

Technology support in cultivation and other practices helped the people in quick disposal of their tasks. Tube well irrigation brings revolutionary change in irrigation system as ground water drawl became easier. Cultivators started spreading irrigation facilities for meeting the agriculture produce for growing population. Similarly tube-wells and hand pumps helped in providing drinking water in all the localities for benefiting the people. The development process through technology support eased the problems of the people to significant extent but ground water resources are very scanty in the state. Drawl of ground water at very quick rate and in large quantity created serious problem of ground water depletion. Now most of the state is suffering from low availability of ground water as drawl process could not match with the resources recovery during monsoon. This imbalanced total ground water system and most part of the state has been converted in dark zone or critical condition where further drawl of water is not permissible. Land fertility has been seriously affected and harvesting rainwater for drinking and irrigation use has been completely vanished.

4.3.5 Atmospheric Changes

Extremity of temperature is the impact of moisture loss controlled by forests. This process resulted in extinction of moisture in sub-surface soil strata and evaporation process intensified. Loss of humidity from atmosphere disrupted monsoon cycle on one side and loss of vegetation cover from the earth surface as resultant impact. Temperature, humidity and evaporation process are natural factors and imbalance in between them disturbed the monsoon cycle to great extent. Presence of dust in atmosphere prevents monsoon clouds to proceed in the direction and low occurrence of rains in western region is becoming regular phenomena as a result of human interface. Human acts of destructive nature remained uninterrupted creating adversities in the monsoon and imbalancing the natural factors in positive role.

4.3.6 Processes of Drought

Processes of drought start from disruption of monsoon and first impact is visualized on crop damage. For cultivation practices, arrival of timely monsoon, regular spells and timely without withdrawal are the process helping cultivation practices. Every process of rainfall is helpful in germination, growth, maturing and harvesting of crops and any disruption at any level is visualized in terms of crop damage. Thus rainfall spells and crops are inversely related and crop damage are treated as dispersal from the regulated system. Monsoon has become erratic and uncertain causing serious problem in cultivation practices as monsoon crops are grown in wider areas and help people to arrange their livelihood and other basic needs which are to be fulfilled through harvested crops.

Drought process starts from crop damage and extends further to drinking water problems. Post monsoon crop is solely dependent on availability of water and in case of monsoon failure, problem of drinking water and expectations of post monsoon crops become dim. Cultivation practices cannot be undertaken but arrangement of drinking water is to be managed from all possible measures to prevent human and livestock survival. Loss of vegetation from earth surface is the resultant impact as total vegetation cover and grown trees, bushes and shrubs are completely damaged for meeting livestock requirements.

Drought is disastrous feature and its intensity is viewed from the impact over the biological factors. In case of most severity of drought, various diseases start spreading from malnutrition or food shortage. The intensity of drought is enhanced through its devastating role and various problems cropped up creating serious or unmanageable conditions. Short term impact of drought are limited to the duration of next monsoon while long term impact are serious on environment, ecology and ecosystem take very vast time to recover. Damage over earth surface is not easily recoverable by ensuing monsoon as intensity of drought creates long term damages, which are to be faced by the people for further duration as per its consequences.

4.4 NATURAL FACTORS CAUSING DROUGHTS

Climate of Rajasthan state varies from hyper arid to sub-humid in different climatic zones. In the west of the Aravalli range, the climate is characterized by low rainfall with erratic distribution, extremes of diurnal and annual temperatures, low humidity and high wind velocity. The climate is semi arid to sub-humid in the east of the Aravalli range, characterized by more or less same extremes in temperatures but relatively lower wind velocity and high humidity with better rainfall. The entire state is characterized by hyper-thermo conditions.

The annual rainfall in the state varies significantly and general trend of Isohyets is from north-west to south-east. There is very rapid and marked decrease in rainfall west of Aravalli range making western Rajasthan the most arid part. The average annual rainfall in this part ranges from less than 10 cms in north-west part of Jaisalmer, which is lowest in the state, to 20 to 30 cms in Ganganagar, Bikaner and Baarmer regions, 30 to 40 cms in Nagaur, Jodhpur, Churu and Jalore regions and more than 40 cms in Sikar, Jhunjhunu and Pali regions and along the western fringes of the Aravalli range. On the eastern side of the Aravalli range, the rainfall ranges from 55 cm in Ajmer to 84 cm in Jhalawar. In plains of Banswara and Dungarpur districts annual rainfall ranges from 84 to 95 cm, while Mount Abu receives highest rainfall of 163 cms, which is highest in the state. Yearly total rainfall is highly variable at different places and most erratic in the western half with frequent spells of drought, punctuated occasionally by heavy downpour in some years, associated with passing low pressure systems over the region.

Southern monsoon, which has its beginning in the last week of June in the eastern parts, lasts till mid-September. Pre-monsoon showers begin towards the middle of June and the post monsoon rains occasionally occur in October. In the winter season also, there is little rainfall occasionally associated with the passing western distribution over the region. In most places, highest normal monthly rainfall occurs during July and August. The number of rainy days during June to October varies widely in different places ranging from 10 in Jaisalmer to 40 in Jhalawar and 48 in Mount Abu. Rainfall during the rest of the period, rainy days range from 2.1 cm in Jaisalmer to 7.2 cm in Jaipur, distributed over 2.5 to 6 rainy days. Basic cause of drought is significant deviation from normal rainfall.

Natural factors of droughts start from pre-monsoon estimates with the development of atmospheric features over the sea where monsoon develop in the Indian Ocean. Arrival of monsoon in Andaman Islands is mid-May and it reaches to Kerala Coast in May end. Further forecasts are based on the advancement process and likely date for appearing into Rajasthan is estimated. Usually monsoon rains start in southern Rajasthan in third week of June in normal conditions and activated in total state in last week of June or first week of July. The first spell of rainfall is quite critical as cultivators grow crop-seeds after occurrence, though earlier field preparations are carried out in June. In case of delayed monsoon by a fortnight to one month, cultivators prefer to grow low duration crops.

Further spells of rainfall help in germination and seeds followed by strengthening of crops and seeding grains. In case of failure of further rainfall spells, seeds in process of germination to growth is disrupted fully or partly depending on the atmospheric conditions. Concluding spell of monsoon in mid-September is helpful in crop yield and its failure damages the crop as per intensity. Similarly drinking water problem is adjudged with the charging level of rain water inside the earth surface where slow rains are helpful in charging the water to ground water resources. Heavy downpour in various areas helps in carrying to established or shallow drains.

Rainfall failure is natural feature governed by atmospheric conditions and intensity of drought depends on its severity. Even normal rainfall with erratic spells damages the crop but ground water is charged to large extent. The atmospheric features help in turning the temperature low which helps in increase of moisture and humidity. Low temperature in pre-monsoon duration does not attract rain water as per well set monsoon travel system. In western Rajasthan in general and Jaisalmer district in particular dust particles raised in the atmosphere due to wind storm create problem in attracting clouds with wind velocity, which is main feature of low rainfall in this region.

Climatic factors play vital role in creating drought conditions. Extreme high temperature of summer turns the atmospheric heat near earth surface and hot air raises high creating vacuum in the region. Cold wages associated with cloud try to fill the created vacuum at the earliest and result in rains. In case of deviations in the atmospheric features the clouds divert to other regions leaving such regions unattended. These are the natural factors which are also the resultant impact of human interface through loss of vegetative cover. Greenery on the earth surface helps in attracting clouds in such regions but denuded areas are usually left untouched creating drought conditions.

In creating drought conditions, all negative factors play their vital role by diverting clouds to other regions and total state area suffers from less or no precipitation. The natural factors are very rigid and no deviation is possible as human acts have deteriorated the atmosphere to great extent. Unless the negative factors are decreased with human efforts, such conditions are bound to happen. Drought conditions are severe when all negative atmospheric features combine together and in case of deviation in all the responsible factors, rainfall occurs to some extent. Accordingly, the monsoon forecasts through the changes in sea pattern and generation of clouds help in rainfall but the directions in which the clouds are moving suddenly change their direction with instantly developed feature mould the cloud direction to other areas and such forecasts are falsified.

The atmospheric changes play vital role in rainfall and low pressure zones created due to atmospheric conditions, suddenly disturb total system and clouds turn their position in other and unexpected directions. These factors are very sudden and known when they create havoc in other regions. Even Rajasthan receives heavy downpours with unexpected atmospheric changes, which have never expected. With the help of satellite, it is possible to make assessment for a day or two provided the atmospheric changes do not disturb the direction. The cyclonic impacts are usually expected for moving in any set directions but sudden deviations change total scenario disrupting the forecasts.

In case of drought, its prior estimation are possible to some extent but such presumptions are not true in terms of intensity and extent of droughts. This is the reason why the drought intensity and extent varies from region to region and further deviation in any one specific region. Since cloud behaviour remains very uncertain through out the monsoon season, real position is assessed in each area and locality in terms of intensity and impact of droughts. Even the severity of drought is unable to be assessed in advance and similarly the conversion of drought into famine conditions is not a matter of forecasts. Atmospheric features do not prove worthy at micro levels and this is the reason of varied intensity of drought even in one region or zone.

Drought disaster varies in terms of area and intensity. There are few pockets in Rajasthan, which are least affected with drought impact. There are some orans where vegetation cover is dense. One such oran is spread over in 40,000 hectares in Barmer district. Jsisalmer, Jalore and Pali districts have also orans of lesser spread, where tree felling is restricted by community convention. Even outsiders could not gather courage to damage such areas. Plantation activities are performed by the local community where each family participates as their ethnic responsibility. Similarly there are some pastures, which help the surrounding areas in the time of severe scarcity, which reveals that plantation helps in checking the adversities of drought to vital extent.

Natural factors creating drought conditions can be mitigated with human efforts to great extent as human interface is responsible in creating the adversities of natural factors. This is certainly true that natural factors creating drought conditions change in long time but such changes persists for longer duration. With this preposition human efforts can succeed in decreasing the intensity and magnitude of drought in a decade but devastating factors need to be curbed strictly and this is the measure to restrict droughts from converted into famine in the state.

4.5 INDICATORS OF MEASUREMENT OF DROUGHT

Indicators for measurement of drought include manual and satellite system. Satellites are able to provide quick assessment of drought severity. Space technology is well developed to carry out assessment of one square metre area in systematic manner where crop damage and ground water, surface water, level of rivers and streams is able to assess. However manual system is deemed proper for such assessment, which help in collection of total information in one month including preparation of reports. Manual system of drought impact assessment is preferred for various reasons. These assessments are carried out in each village with the help of district revenue administration and complete assessment is submitted to district administration within two weeks and total state assessment is possible in one month. These assessments are made for crop damage and drinking water conditions in each village.

4.5.1 Indicators of Measurement of Crop Damage

Indicators of crop damage are measured in each field by the revenue agency of district administration. Lowest person in this ladder is Patwari associated to Girdawari circle of district. One patwari is responsible for all revenue activities including collection of land revenue in four to ten villages depending upon the area spread and population. In very big village, one patwari looks after work of one village while others are responsible for maximum ten villages. In case of drought conditions, District Collector calls all the Girdawars of district to instruct assessment of crop damage. This system is well defined as drought is recurrent feature and revenue administration is well accustomed with the system. Girdawars of each girdawari circle are required to call all patwaries for communicating similar instructions along with time requirement of submission of the information. Patwari is required to villages of charge to make assessment of drought which is performed by eye-view system visiting cultivated fields.

Indicators of measurement of crop damage are divided into four categories, where crop damage assessment is done in terms of less than 25 percent, between 25 to less than fifty percent, between 50 to less than 75 percent and between 75 to 100 percent. Such assessments are usually made in the month of September, when such assessments are possible when the crop is standing in fields and crop assessment is possible. Patwaries are well versed with the system and assess crop damage in few hours viewing most of the fields. Patwaries are required to remain unbiased in their assessment and carry out paper work during night in view of importance of the task. Such reports are submitted to Girdawar in the specified time frame. On receipt of information from all the Patwaries, Girdawars consolidate reports in terms of each village indicating crop damage in requisite terms. Revenue agency of district compiles information of total district and sent to state revenue department for compilation of information of total state.

District reports compiled at state level reflect number of villages affected with crop damage of more than 50 to 74 percent and from 75 to 100 percent in village numbers, affected human and livestock population, crop damage in rupee terms and recommendation of District Collector for suspension of revenue collection in case the crop damage is more than 50 percent. Fifty percent crop damage is the dividing line in crop damage as less than fifty percent crop damage is treated as normal feature. State information consolidated in this preposition is submitted before the government for taking appropriate decision. Further process is also well defined as in case of large spread of area affected from drought, proposal is required to be submitted to the Government of India for providing assistance of requisite magnitude.

4.5.2 Drinking Water Situation

Such assessments are made by Patwari in each village by assessment of level of water in wells. In case of piped water supply schemes, such assessments are made by Public Health Engineering Department in each district. District Collector incorporates the position in the report along with the recommendation for severe drinking water problem in villages where existing water supply position is expected to become deteriorated in coming months. This component becomes part of the state report in case some assistance is required from the Government of India for meeting drinking water arrangement as crash programme in the duration till next monsoon. Drinking water issue is related to social responsibility of the state government to avoid unpleasant situation during the scarcity conditions.

Manual process of drought measurement in terms of intensity and magnitude is preferred in view of the fact that there remains some flexibility to expose position to the extent which permits assistance from the Government of India. This is an issue of centre-state relationship where each state try to expose higher intensity and demand high quantum of funds for meeting all the requirements. On receipt of such reports in the Government of India, central team is formed to have first hand assessment and the recommendation of the central team are helpful in finalization of quantum of assistance. Usually each state tries to expose the position to get higher assistance and the central team usually recommends lower amount to assist the state for facing the challenges of drought.

Flexibility in manual process is easier to twist the situation exposing as severe drought condition where the Government of India, provides funds for initiating wage employment in affected regions. Money is received in terms of food grain and cash to face the challenges in view of aggrieved situation. Drought is associated with human factor and state government is required to assist people in distress. Drought measurement indicators help in exposition of severe condition and insist the Government of India through administrative and political channels.

In case of measurement of drought through agreed indicators is well versed to the state government and the central government and every state government try to impress upon the central government by exposing the severity of drought. The process of drought measurement is tried to remain flexible as state government is also sure that the central government is not going to sanction as per their estimates. The total system is based on mutual distrust understanding the role being played in this system. The ultimate responsibility of facing the famine condition is of the state government, which tries to dovetail sectoral funds to various activities where such arrangement is feasible.

Space technology based assessment of drought conditions is also not free from judgement in proper manner. Usually sowing season remain varied even at local level as some cultivator prefer early sowing while others prefer to wait for some time to assess monsoon trend. Therefore crops of different types cannot be adjudged with aerial system of assessment as such judgements cannot be true in total cropping pattern. Various crops have their set growth pattern and no universal pattern of judgement of the crop damage is possible. Furthermore, these indicators of crop damage assessment are applied in all the states and manual system is helpful in assessment of actual position, if administrative and political biasness is kept aside.

The system of assessment of crop damage is itself biased as crop damage up to 50 percent is treated as normal phenomena and no assistance is permissible in such conditions. This system is based on the analogy that the crop damage up to 50 percent need to be borne by the cultivators. From economic and social point of view this pattern has lacunae as cultivators use inputs in crop growing and there is labour component of cultivator or support of mechanical devices in case used. Both the components are equal to cost of harvested crop. In case of 50 percent damage is left out from crop damage pattern, this cannot be treated as justified. Even after application of technology support, most of the cultivation practices are still done through conventional practices. Cultivators are engaged in this activity simply for the reason that they have no alternative for their livelihood. In case of crop damage, total duration up to next monsoon remains at the stake and this duration is highest in single cropped areas and also in case of marginal and small cultivators.

From economic point of view, even the wage component of cultivator is not received in the condition of Rajasthan state where monsoon is most erratic and uncertain. About 80 percent of cultivable lands of the state are single cropped area which is based on the mercy of monsoon. In case of crop damage up to 50 percent, no assistance in the form of wage employment or deferment of land revenue is possible whereas the expenditure incurred as inputs for cultivation are hardly met by harvested crop with damage of 50 percent. This is the reason why cultivation practices in the state are means of survival base rather than commercial activities in other developed states of country where monsoon behaviour is not so erratic or uncertain.

5

Occurrence of Droughts and Socio-Economic Consequences

Drought is most serious natural disaster specially in case of Rajasthan state where total region is categorized as hyper-arid, arid, semi-arid and sub-humid categories. The vulnerability of drought has reached to very serious conditions, where even small variation in occurrence in rainfall creates very harsh consequences. In addition of crop damage of varied nature, depleting ground water resources and dried water bodies preserved from centuries. The western Rajasthan was severe drought prone area from long past but southern and eastern Rajasthan has also become prone to recurrent droughts due to human interface. Drought affects common man in various manners. Cultivators face serious crop damage, while all the people of rural areas are compelled to bear drinking water and irrigation problems as resultant impact of low rainfall. Water resources are not more than one percent of country water resources, where total state area is 10.41 percent and 5.50 percent population of the country. This reveals the problem of water resources from all known and accountable sources.

Drought conditions create serious water crisis for all types of requirements specially in rural areas. So far as urban areas are concerned, state government tries its best to manage the situation even by crushing the interest of cultivators from water bodies used for irrigation from long past. The national Commission on Agriculture in its report on 1976 has classified western Rajasthan as chronically drought affected area due to prevalence of drought frequency over 40 percent of years. But position of other regions has been deteriorated to similar extent. Western Rajasthan with 61 percent of area and 39 percent of population is drought prone due to paucity of water while remaining part with 39 percent of area and 61 percent population is drought prone for extreme pressure of population and natural resources have been vanished for uncontrolled human interface with the nature.

Dominance of widespread aridity and truant climates with monsoon fiats are most critical factors of drought which frequently converted into famine in view of imbalanced environment and ecology factors. Extensive deforestation over centuries has altered hydrologic cycle and tended to increase the incidence of droughts in every part of the state. Droughts create acute deficiency of food, fodder and water due to low rainfall. These commodities are managed by the state government to provide relief to the people, but crop damage vanishes the purchasing power of the people to arrange essential commodities.

5.1 HISTORICAL TRACE UP OF DROUGHTS

The hydro-meteorologists, hydrologists and agronomists viewed the droughts from their respective angles. The information available of past drought conditions are limited to districts covered with droughts, population affected and deaths of human and livestock population occurred as a result of drought. Three successive Famine Commissions were constituted during 1878-1900, where 1878 commission recommended employment on large work of permanent utility, cooked food for disabled persons and circulated a model Famine Code. The Second Commission of 1889 adhered to the earlier recommendations and also added necessity for inclusion of district wise works and positive policy for suspension and remission of land revenue. The Third Famine Commission recommended to revise famine code and new irrigation policy for developed irrigation facilities for combating droughts. The Irrigation Commission of 1903 recommended for extension of irrigation facilities as alternative of rainfall failure and also the miserable condition of poor.

The records of incidence of past droughts and famines are based on the revenue records revealed through marked prices of commodities, migration of people, human and livestock discern. Records of last fifteen famines are traced in the duration of 1792 to 1899 and there are reasons to believe that the history of Indian arid zone remained full of drought stresses in view of following conditions:

(a) Nine times out of ten in the past associated less problematic areas with similar calamities.

(b) There was large scale migration during the famine years and no person was able to traced to be traced to reveal land details.

(c) Large scale human and livestock deaths compelled people to learn lesion for water conservation in scarcity conditions.

(d) Livestock conservation with migratory trends developed in the duration for retaining purchasing power to face droughts.

The chronicle droughts in the historical archives in Indian arid zone are open from 1792 when there was widespread famine due to failure of monsoon in almost of the state, which caused severe crop damage, drinking water shortage, massacre human and livestock casualties as most of the ground water resources were depleted and surface water was extinct. After 12 years, there was devastating drought in 1804 spread throughout central India, where Rajasthan was seriously affected. The drought of 1812-13 occurred after eight years due to failure of last spell where food, fodder and water remained in the reach of the people. Drought of 1833-34 was most hazardous followed by drought in 1838-39 and human population suffered very seriously. Drought of 1848-48, 1850-51 and 1853-54 were quite close with little replenishment from earlier devastation. These years remain worst in the history of Rajasthan as people suffered with human and livestock loss to great extent.

Droughts of 1868, 1877, 1891-92, 1895-96, 1898-99 and 1899-1900 were most severe in terms of crop damage and people were starved for food and water. Migration of human and livestock to water bodies became regular feature. Loss of human life was severe and people remained at their homes were compelled to eat tree leaves, roots of trees and even wild animals for their survival. Such droughts proved Malthusian theory of preventive measures of population control as pressure on earth surface became unbearable due to destruction of vegetative cover. In the regime of princely states, the rulers were aware of the drought consequences but unable to provide significant relief to the people. 75 percent of livestock was perished and human loss was not less than 30 percent. Most sufferers were the small cultivators and wage earners, who had no work to attain their livelihood.

Impact of drought remained severe in pre-1900 period as there were very low infrastructure network in the state. Roads were constructed as per state importance and basic amenities were quite rare. Mostly people were supposed to care for their necessities and state role was little welfare but land revenue collection was quite strict being important source of princely states. However, large construction activities were carried out as per available records but at limited sites. Intensity of drought is associated with drought proofing measures. Usually wealthy people used to construct water bodies to help people and wild life and such practices are common in the state. Digging well was treated as ethnic cause and rich people were vigilant for their responsibilities. Artificial lakes and reservoirs were generally constructed for survival of wild life for hunting.

In the first half of 20th century and during 1901 to 1950 there were twenty worst years in the history of Rajasthan where people suffered quite severely. During drought years, prices of essential commodities remained uncontrolled as no such measures were practiced in rural areas. Furthermore people had no voice to express their problems and such devastating acts compelled people to treat such calamities as their ill fete. Communication measures were limited and drought like conditions become recurrent due to constant deterioration in the environment and ecology. People used to preserve rain water for meeting their drinking requirements and water was treated as precious commodity. People reside in rural areas near water bodies or their ability to harvest rainwater. Jaisalmer and Barmer districts still have very little village culture.

After inception of planned era, the intensity and magnitude remained recorded as state government treated its social responsibility to make proper assessment of drought conditions and helping the affected people to best of their ability. In the duration of 1950-51 to 2002-03, out of 53 years only 6 six years were drought free in whole of the state, i.e., 1959-60, 1973-74, 1975-76, 1976-77, 1990-91 and 1994-95 where no part of the district was affected with the drought disaster. During 27 years the intensity of drought remained very severe as more than 20 districts of the state were gripped with drought with wide spread impact. During 13 years 11 to 20 districts remained affected with the problem of drought and during 7 years the drought phenomena. During 10 years all the districts of the state were affected with drought. Similarly during 18 years, more than 50 percent of the state villages was affected with the disastrous conditions witnessing the intensity and magnitude of drought.

5.2 FREQUENCYAND NATURE OF DROUGHTS AND FAMINES

Intensity of drought is resultant impact of rainfall, which is the base of monsoon cultivation. Timely rainfall with suitable margin in various spells is helpful in cultivation practices. Untimely and heavy downpour in single spell is also dangerous for crops. Therefore cropping pattern require smooth rains at the requisite and cultivators try to match their crops with the occurrence of monsoon. However, cropping process remains uncertain till harvesting and processing. Meteorological drought is termed with quantum of rains in monsoon duration and also in the total year. Crop damage is also associated with this pattern but not completely co-terminus.

Drought is termed with the occurrence of normal annual rainfall and in case of deficiency up to less than 25 percent, treated as mild drought. Rainfall deficiency ranging between 25 to 50 percent is treated as normal drought. When the rainfall deficiency becomes more than 50 percent but less than 75 the intensity of drought is termed as severe. When the rainfall deficiency is more than 75 percent and up to 100 percent of normal annual rainfall, the condition is most severe drought and there is every fear of converting into famine.

Usually drought and famine are termed in synonymous terms but have very significant difference in terms of impact and magnitude. Drought is normal natural phenomena happens in every part of the world but creation of famine conditions depends mostly on process of relief measures to save the sufferers from starvation. Meteorological drought is treated when failure of rainfall is more than 75 percent, if earlier was normal rainfall year or rainfall deficiency up to 25 percent. The agricultural drought is treated on crop damage more than 50 percent, as less than 50 percent crop damage is treated as within the bearing capacity of the cultivators. Hydrological drought is worst situation where water bodies are completely vanished.

Drought is treated as natural phenomena but if relief measures are not initiated in time, people start suffering from the devastation of drought and famine conditions prevail over the affected area. The harvesting time is the period when the cultivator expects crop, as its most hopes are concentrated on the harvested crop. Cultivators postpone their basic needs till grains are received and cultivators sell their surplus to meet other requirements. In case of agriculture labour, usually payments are made on sale of grains. Even the cultivators borrow commodities of daily requirements to be paid after crop. This is regular feature of transactions with the cultivators and businessmen also hope for their earlier recoveries.

Similarly, market also hope transaction of agriculture produce and purchase in bulk to sell people throughout the year. In case of crop damage, cultivators are not the only sufferers but total market is affected with this scarcity. The state and national economy is also get serious setback in terms of loss of net domestic products or gross domestic product. Various processing units suffer with this situation as crop failure is shock to many other persons associated with cultivation practices. In case of most severe drought, whole economy of country suffers vitally.

Frequency of drought varies from region to region in the state. Jaisalmer district is most critical from frequency point of view as every drought is bound to affect this district in view of most adverse conditions. During last century Jaisalmer district had 67 drought years where 42 years were moderate drought and 42 years had severe and most severe droughts. Barmer district is just adjoining to Jaisalmer but had 47 drought years where 18 years were moderate droughts and 29 years with severe and most severe droughts. Bikaner district is third in sequence having 45 drought years where 23 years had severe to most severe droughts. Jodhpur district is next district having drought frequency of 42 years where 16 years were severe to most severe drought conditions.

Jaisalmer district is most problematic region of the state covering 11.22 percent of state geographical area but normal annual rainfall is least with 18.55 cm. Even in normal conditions, the district have various constraints reflected through least density of population of 13 persons per sq km in comparison to state figure of 165. Wasteland area is more than 67 percent of district geographical area covering most of the western part. Canal irrigation system has created some hope in the people for better prospects but it will develop gradually.

5.2.1 Frequency of Droughts

Droughts have been categorized into moderate drought, severe drought and most severe drought. Moderate drought is the situation where the rainfall occurrence varies 30 to 50 percent less than normal rainfall. Annual rainfall is the situation, which is one of components of crop failure. Even if the rainfall with indicated level below normal level may not affect the crop to same extent as annual rainfall occurrence may remain meeting all the requirement of agriculture sector. Similarly the crop damage is possible even the occurrence of rainfall remained just equal to normal rainfall but the rainfall spells may not be meeting the cultivation at the time of germination, crop survival and seeding before harvesting time. There are multiple factors causing normal, deficient and excess rainfall. Rainfall is the first such factors for crop damage to moderate to most severe level. Even the crops can be successfully grown in low rainfall conditions provided the beginning and further spells are in right time. Frequency of drought is result of various associated factors. Crop failure of moderate frequency is treated as general crop damage below 50 percent though this is general observation but actual assessment is done from each village to know the actual position of crop damage.

Frequency of drought is severe on rainfall failure of significant extent. The crop damage position is assessed in each village by visiting each cultivated field and this process is required to make estimates of crop damage in money terms. Even in severe drought conditions, crop damage remains varied in different regions and sub-regions depending upon dryness in the atmosphere and moisture loss. Severe drought owing to failure of subsequent spells of rainfall or early retreat of monsoon. Frequency of drought remains varied as in some cases, grain loss becomes severe but some fodder is possible.

Most severe drought is the result of serious rainfall failure and affects large area in terms of spread and frequency. Even in most severe drought conditions, there can be very few exceptions frequency of drought is not that severe and people get some crop but such cases remain very few as impact of drought remains widespread due to atmospheric conditions. Areas with large vegetation can avail such variation as frequency of drought remains slightly low drastic than large area. There are some pasture protected by village community and such villages do not fact vital hardship of nature. Areas in vicinity of rainy drainage system are able to retain some moisture where intensity of drought has low frequency of drought. In view of dryness in the atmosphere, even small deviation in rainfall create serious problem in crop conditions.

5.2.2 Nature of Drought

Rajasthan economy is basically agrarian with most part of the state dependent on monsoon cultivation. About half of the cultivated land has become degraded due to salinity and alkalinity. Thus the cultivated area is regularly squeezing in view of long dryness. The nature of drought is most varied and there are very few years of good rainfall. Barren hills and sand dunes are covered with occurrence of precipitation but vegetation loss is eminent before summer. Erratic nature of monsoon is not dependable for cropping and cultivators with rain cropped suffer significantly. As a supplemental source of livelihood, people keep livestock in large number, which graze in open completely, vanish the little vegetation grown over the earth surface.

In the situation of monsoon failure, drought condition prevail over the affected area. The intensity and magnitude is dependent on the nature of rainfall. When the rainfall failure damages crop to significant extent, surface water vanishes due to evaporation and ground water depletion is severe, the nature of drought becomes hazardous. This situation is combination of agriculture, hydrology and meteorological factors with severe intensity. The situation of monsoon changes quite rapidly and monsoon failure is feared at each spell. The geographical conditions are also not favourable to this region of the country. There are some drought prone areas where occurrence of drought is eminent. In every drought year, Jaisalmer, Bikaner, Barmer and Jodhpur are the critical areas affected most irrespective of the area spread of drought.

The environment, ecology and eco-system of the state has been deteriorated to great extent. Cultivated area is being degraded losing its productivity. Ground water resources have reached to most critical stage where extinction of these resources is feared in near future if precautionary and corrective measures are not initiated. Human interface is the main cause for creating all the adverse conditions in the state. People of the state never think the future consequences of their acts. Even the state government remains as silent spectator of all devastating acts creating critical conditions. The vegetative cover is almost lost and there is no one to feel the responsibility for restoration of environment and ecology of the state.

Every drought is dangerous for the soil productivity and reclamation process require some time to restore damaging acts of drought but when there is no process of reclamation, the nature of drought becomes further aggravating. In the present situation, all the mitigation measures for drought proofing could not yield desired success but the devastating acts are spreading continuously with increasing magnitude.

5.2.3 Drought and Famine Conditions

Drought is natural condition occur due to failure of monsoon. Crop damage, water scarcity and increase in dryness are the resultant impacts. People of rural areas get setback in terms of their social and economic conditions. Most sufferers are the small and marginal cultivators, agricultural labours, who loose their hope for return from cultivation practices. Rural economy is almost dwindled with the severity of drought, specially in the situation of complete crop damage, acute water shortage and loss of vegetative cover. There becomes acute food shortage to rural people, who expected crop yield for meeting their annual requirements. Fodder shortage and vegetation loss create miserable condition for survival of livestock. Drinking water shortage for human and livestock can be dangerous for large scale loss of life.

After effects of drought are most critical as people dependent on cultivation practice lose their purchasing power even for meeting their most essential requirements and moneylenders helping in the hope of crop yield and recovery, withdraw further credit in view of uncertain situation of payment of earlier and new purchases. In this condition, cultivators and other persons dependent on agriculture are put to very critical condition, where their survival remains at stake. Combination of all such situations is the origin of famine conditions. Though drought and famine are the resultant impact of monsoon failure, but drought is the event of natural acts while famine is the creation of condition Though drought and famine are the resultant impact of monsoon failure, but drought is the event of natural acts while famine is the creation of adverse conditions and delayed relief measures.

Crop harvesting time of monsoon crops is usually October, while the crop damage and water shortage conditions are known by September end. State government prepares memorandum after compilation of district reports and submits the position before the cabinet and later forward case to Government of India for assistance as per natural calamity relief procedure. The total process of sanction of central assistance takes time and relief measures are possible by December or January. Sometimes this period becomes unbearable for affected population and famine conditions start cropping up in most dangerous form where people and livestock suffering intensify most. Even starting relief measures and availability of assistance takes more time.

Under relief measures, wage oriented works are started in critical areas and wages are given in cash and food grain to help the affected people. Fodder depots are established in the affected areas for survival of livestock. Medical teams are sent to problematic areas for survival of people from endemic and epidemic diseases. Total district administration is alerted to manage every short of requisite relief. The government machinery tries its best to avoid human and livestock as moral commitment and take all possible measures to decrease the intensity of famine conditions. If there is undue delay in relief measures, the problems of affected population start intensify their magnitude and there are large scale loss of human and livestock life. In drought affected areas, various seasonal diseases create very unpleasant situation and in case of delay in quick remedial measures, position become out of the control of the administration. Electronic and print media exaggerate the position to expose the sufferings of the people.

Fighting with famine conditions is very typical issue as district administration has to face challenging conditions. Any delay in timely assistance of food, fodder and water aggravates the famine conditions and loss of human and livestock in large number creates very unpleasant situation for the government machinery. Success of the government machinery depends on timely availability of requisite assistance, where District Collector intends help of various types as resultant impact of famine conditions. Timely assistance to affected areas is possible with proper coordination with state level departments as the district level respective units are unable to control total situation.

Sometimes famine conditions aggravate very fast in wide spread area and its control becomes very difficult. Very slight delay in timely relief measures results into large scale casualties. Successful handling of famine conditions is lease loss to human and livestock. Affected people forget their individual losses and engage them in relief works to manage their livelihood. Drought and famine conditions prevail usually up to occurrence of monsoon and district administration draws its combating plan up to June. In case of further rainfall failure, the famine conditions become very severe and control of such situation becomes very difficult. Impact of drought and famine takes long time in recovery even when the drought and famine are active during one year. In case of repeated problem, the problem is critically exposed to dangerous level.

Disaster management is special aspect of administrative functioning and its performance is adjudged by minimizing losses to most possible extent. Now drought and famine are well accustomed to district administration and take advance action to overcome problems in case of monsoon failure. Droughts and famines have multi-faceted problems in terms of intensity and magnitude and new series of problems aggravate the problem further. Working over unknown problems become difficult, as the administration has no past experience. New challenges with no past records create problems without understanding impact and intensity in aggravated conditions. There are instances, when district administration become stagnant The natural conditions of environment and ecology are deteriorated to most critical extent due to uncontrolled human interface. Natural fertility of soil has been deteriorated to vital extent due to large scale deforestation. Wind erosion is very critical in most of the deforestation affected regions. Long dryness of the atmosphere has further aggravated the conditions and slight deviation in rainfall creates severe devastation.

Ground water depletion has reached most dangerous level and some areas have extinct ground water resources. In such regions, only rainfall is the source of cultivation. Even failure of one spell of monsoon create drought and famine conditions of serious extent. Jhunjhunu, Nagaur and Sikar districts have many such critical areas, which are compelled to face serious famine challenges due to human interface. Out of 237 blocks in the state, ground water position has become most critical and about 200 blocks have been converted into dark zone and critical zones, where further draft of ground water is most dangerous. Even after knowing such dangerous situation, draft of ground water is uncontrolled as people and government both are indifferent with the serious consequences of the situation.

Occurrence of drought and famine conditions and initiating relief measures is not the permanent treatment of the problem as every drought causes further intensified conditions and it is difficult to make proper assessment of deterioration within next two decades. Retreat duration of drought is quite long in case of state wide coverage. Situation of Rajasthan is different from other states as there are few perennial rivers and very scanty ground water resources, which have reached to most critical stage. In such conditions, occurrence of most severe droughts has become recurrent feature.

5.3 DROUGHTS AND POPULATION

Droughts affect population as per its intensity and magnitude. In case of state wide drought most of the rural population is affected to severe extent. Droughts are the result of natural factors and human interface further intensifies the problem. Impact of drought lies over the population in many ways. Crop damage is not only problem of cultivators but total country is affected with the deficiency of agricultural produce. This is true that drought in one region is bearable to the country as other areas balance the total damage, but the process of development is seriously affected. Increase of population at fast rate imbalances natural factors and pressure on land increased to great extent. In the state there are regions having more than 40 percent decadal population growth rate and arranging agriculture produce for fast increasing population becomes difficult specially in drought condition. In the state of Rajasthan, drought and famine hamper process of development as production in good monsoon year becomes setback to ensuing crop production. Average crop production in Rajasthan is stagnant from last three decades. Population is affected by drought in many ways as large section of population is directly affected by drought while some are indirectly affected.

5.3.1 Population Directly Affected with Drought

Affect of drought remains varied over the population directly suffer from drought conditions. Most sufferers are the marginal and small cultivators as they have very meagre cultivated land mostly unirrigated, where monsoon crop is possible. Marginal cultivators own less than one hectare land, while small farmers possess cultivable land between 1-2 hectares. Their lands are marginal and degraded where crop yield in general conditions is quite less than the average produce. Semi-medium cultivators having cultivable land ranging between 2-4 hectares suffer from drought if their land holdings are unirrigated. Cultivators owning operational holdings more than 4 hectares suffer less as such cultivators can manage irrigation facilities and they are able to compensate their produce in post monsoon duration.

Agricultural labours are also sufferer with drought as they cease to get employment in view of crop damage. In many cases, their wages are deferred to further duration as cultivators are not in position to provide wages due to crop damage. Small and marginal cultivators wholly depend on the crop produce and sell their surplus produce in the market to purchase their other essential commodities. In most of the cases, such cultivators purchase essential material on deferred payment basis and such payments are usually made from sell of their produce in the market. Even in normal conditions, such cultivators are compelled to sell their surplus produce to the village person where the price of their produce is settled at much lower rate than in the market. Such cultivators have no means to carry their produce to regulated market and get lesser price from the person, selling their commodities throughout the year. In case of crop damage, such persons continue to sell their commodities to cultivators on deferred payment basis and cultivators have no option to sell their produce to any other person as part of obligation.

Purchasing power of the cultivators become stagnant in case of crop damage and such cultivators feel helpless with their miserable conditions. The affected people have one complete year, when they expect crop arrival in case of good rainfall. The sufferings of such cultivators are most pity in case of any unexpected expenditure due to disease or death. The extra burden is usually met by moneylenders who provide money at high rate of interest and mortgage their land or house. In case of non-payment of dues, the cultivator become labour for arranging livelihood through wage earning. Most of the wage earners are ill feted to surrender their land or house in view of their helplessness to repay loans. Droughts are very ill feted to such people.

5.3.2 Indirect Affected Population from Drought

Indirect affected population from the drought conditions is also varied from village level shop-keepers to district level grain purchasers. In case of crop damage, their business is seriously affected as they solely depend on the bulk purchase and sell such commodities throughout the year. Other affected people from drought are sellers of various commodities, required by villagers. Their business is also affected as small cultivators lose their purchasing power completely or significantly. The state is also affected with decreased crop production.

Small cultivators are almost half of total cultivator community. Even other cultivators get setback and market is completely dwindled with crop damage. In such condition, businessmen purchase commodities from other areas at higher cost and number of purchasers are decreased to great extent. Loss of purchasing power become stagnant in business and commodities are available at higher price. Such condition is matter of serious concern to the lower and middle class people of the society. Steep rise in the cost of commodities affect on the purchasing power even for essential commodities. Droughts are irreversible loss to the people of the state as large population is seriously affected irrespective of their urban and rural settlement.

5.4 DROUGHT AND SOCIAL CHANGES

Droughts are serious disaster for the society as a whole as social activities get setback through the consequences of drought. The poverty line is the cut-off line between poor and non-poor and survey of Below Poverty Line families is carried out once in five years. In this duration there are serious social changes as consequential effect of drought. Droughts and famine conditions are responsible for serious social changes in whole of the state. People possessing assets for their livelihood remain at stake and most of the people are compelled to lose their assets of survival base. In case of poverty criteria, one head of family owning land more than 2 hectares is treated above poverty line. When such land is distributed among dependents, all the individual owners possessing smaller land and surviving on cultivation practices become poor in view of low operational holdings. In case of droughts, such families lose their cultivated land for meeting their emergency requirement in drought conditions and ultimately become landless labours. Retaining cultivable land during drought become unbearable for most of the families. Some persons capable to manage their families and able to generate surplus money purchase such lands and they become better off as impact of drought. Transaction of operational holdings is regular feature of rural society and responsible for serious rise and fall in the social structure.

Usually poor people have very little capacity to face the challenges of droughts and famines. Land is the survival base of most of the rural people and rural society is governed by exploitation of weaker sections by all possible measures. Overcoming from drought conditions is not possible for many of the poor with lesser earning members. Economically weaker families get serious setback specially in case of overdue from banks or moneylenders. Drought conditions dwindles economic structure of many of rural families. Livestock population has been decreased by 10.10 percent during 1997 and 2003. Loss of cross-breed cattle remained 120.05 percent, sheep 31.05 percent and camel 25.61 percent. In this duration, the severity of drought remained very disastrous as in 1997-98 24 districts of the state were affected from drought, in 1998-99 drought affected districts were 20, in 1999-2000 number of districts affected by drought were 26, in 2000-01 the severity was most severe as 31 districts were affected. In the year 2001-02 the drought intensity was rather less as 18 districts were gripped with drought conditions and the drought of 2002-03 was most severe as all the 32 districts suffered by drought and famine conditions.

Livestock rearing is supplemental but dependable source of survival base of most of the rural families. In western Rajasthan livestock population is more than double of rural population where intensity and severity of drought is quite high. Usually poor people rear livestock generally sheep and goat in herd of 100-200 and migrate to other states and return in the month of July. Women, children and old people reside at home for cultivation and their livelihood is mostly dependent on livestock. Even landless families survive on small livestock in view of multiplying very fast. Even milk production is quite significant in the state. Loss of livestock in large number is setback in their occupation, but drought conditions are not under human control.

Social change is also significant in drought conditions as diseases and deaths are prominent social conditions where the affected families have to spend sizeable money on treatment and social rituals. Such emergency requirements are met by moneylenders where the poor families are unable to retain their land and other valuable assets. The serious problem of rural society is neglect of loaning by banks as it is not performing loaning where recovery is difficult. The affected people by drought are left at the mercy of the moneylenders, who exploit such families by high interest rates which multiply with crop damage. There are clear instructions from Reserve Bank of India to allow individual loaning up to Rs. 15000 without any guarantee but banks overlook such practices and setback of development of rural economy. Usually the requirements of rural people for social causes are within such limits, but in absence of easy credit facilities, social changes in rural society are inevitable.

Moneylenders exploit the rural people in view of their helplessness for access of credit from banking and credit institutions and this situation is most critical for the rural society in scarcity conditions. In case banks bear this responsibility, rural society can be saved from serious social sufferings. Usually rural poor defer their marriages in drought years but diseases and deaths are most critical social events dwindling rural economy in general and most critical extent in the drought conditions. Even deaths occur during maternity causing setback as the facilities for treatment remain beyond reach of rural people. Cost of treatment is beyond reach of poor families and conventional practices of child birth are performed mostly in rural areas in distant from such facilities.

In serious drought conditions many people migrate to cities and other states for employment, which puts extra burden of the cities as slum areas develop with this phenomena. In case total family is migrated from village in case of loss of cultivated land and no work during famine year, villages are deserted with large scale migration. Usually rural people have very meagre employment avenues other than cultivation and wage earning Both these activities are ceased during drought and famine. Even the cultivable lands lose their productivity, which compels them to migrate to cities. Though migrated people to cities live in very miserable conditions but able to get employment to survive their families in village or city. Such families try to settle down in cities and their return to home become difficult.

Requirements of rural poor people are bare minimum limited to their most essential commodities as they cannot afford much. During famine conditions, most of the families remain dependent on relief activities and try to avail maximum duration of work up to ensuing monsoon. If work is possible for total drought duration, their sufferings are decreased to vital extent. If relief works are for very limited duration due to low assistance, problem of rural poor are multiplied manifold. Generally relief measures have become essential for rural economy.

5.5 EFFECT OF DROUGHT ON SETTLEMENT

Settlement is part of revenue administration which deals with the ownership of the land of the state. The settlement classification of land is based on use of the land for any specific purposes. According to settlement of village records, every land has its ownership. Area under forests is owned by State Forest and Environment Department. Usually use of forest land is not possible in general conditions but encroachments are made with the connivance of administrative machinery. Figures quoted by Rajasthan Environment and Forests Department claim for possessing 32488.01 sq. kms area under forests comprising of reserve, protected and unclassified forests while settlement record as per village papers nullify their claim by quoting only 26510.14 sq. kms of area, which reveals that 18.40 percent forest land has been encroached by the people. Such encroachments are made for cultivation purpose. Change in ownership of forest land is not possible but this is actual situation of squeezing for forest land. This processing is not the only event and encroachment over forest land is the reality which is authenticated in government record through illicit measures.

Village land is usually left for permanent pastures and tree groves and usually gram panchayat owns such land. Development of such land is the responsibility of Panchayats, which can protect land for pasture development and charge some fees from each user for maintenance. However, such lands have become denuded due to uncontrolled grazing and encroachment over such lands cannot be ruled out. Barren and wastelands are the property of the state, where plantation activities can be carried out and land can be utilized for productive purposes. But such lands are not free from large scale felling of trees by people. Forests, waste and barren lands can be best used for raising vegetative cover over the earth surface and helping in moisture conservation. Encroachment over such lands is also inevitable.

Cultivated land and fallow lands are usually owned by individuals for cultivation purposes. Usually fallow lands are cultivated land but left unattended due to low rain or for being severely degraded. These lands were also forest or pasture lands but encroached in due course of time for meeting cultivation requirements. Fallow lands are generally left unutilized for various reasons viz. regaining soil fertility, land degradation due to salinity and alkalinity etc. In Rajasthan, this role is usually played by famine and drought by decreasing the fertility and in due course of time such lands become waste land on permanent basis. In Rajasthan cultivable land is 207.54 lakh hectares of which 47.93 percent fallow land has became wasteland or barren land with continuous degradation. Therefore the settlement of land under different uses remain on changing as human interface has always disturbed the environment and ecology of the state. Natural factors start expressing their ill effects due to changed atmosphere of the state. However, settlement classification keep on changing due to negligence of the state, district and local administration.

Settlement of the state need addressal of irrigation facilities which remains varied due to impact of drought. The net cultivated area remained varied significantly between 108.07 lakh hectares to 167.65 lakh hectares and increase and decrease in fallow land is also varied due to monsoon failure. Net intensity of crops over the cultivated land remained varied between 121.21 to 133.47 percent which reveals that only 21 to 33 percent land used twice for cultivated use. If percentage of irrigation facilities is analyzed with this position reveals that even irrigated land is not capable to be used twice under cultivation. The drought situation puts serious impact over the settlement pattern of the state.

Area under cultivated has been generated or increased in command area of Indira Gandhi Canal project area as total settlement has been established in new form. The land under waste and barren lands has been converted into cultivated and irrigated lands and that process has added some new area to remain cultivated for centuries. In case of other regions or areas, the cultivated area continues to increase or decrease. The increase in cultivated land is due to encroachments while decrease is the impact of land degradation. Settlement of land remains updated annually as the land use pattern under all the categories is updated each year. The encroachments can be successfully tackled if there is close coordination between forest department, Panchayat Institutions and land owners. Usually encroachments are not taken seriously as there is provision even for imprisonment over such acts.

The settlement of land record can be helpful to other departments in assessment of their respective land ownership as these records are updated on annual basis, where all the details of land use are recorded in case of any deviation from earlier pattern. Cultivation practices and irrigation of crops depend on monsoon spread. Net cultivated area remained varied by more than 50 percent and irrigated area varied by more than 28 percent from lowest level. This variation is significant as first widespread rainfall create hope in the cultivators to use their maximum land under rain fed crops. Decrease in net cultivated area is the resultant impact of disturbance in initial rainfall phase where cultivators prefer to cultivate in high yielding area only. Similarly irrigation practices are significantly affected by rainfall as ground water charge is helpful in higher draft. Canal irrigation is possible in Banswara, Bikaner, Ganganagar, Hanumangarh and Jaisalmer while 13 other districts have some pockets where canal irrigation is possible. Irrigation from wells and tube wells is generally practiced in total state subject to availability of ground water. Tanks, watershed and anicuts have very scanty area under irrigation as per settlement record.

5.6 IMPACT OF DROUGHT ON LAND USE PATTERN

Land use pattern is seriously affected by drought conditions. This change is continuous feature and land degradation is the ultimate situation. Usually forest and pasture land is encroached for meeting cultivation requirements. People have their expectation of maximum coverage of area under agriculture practices to feed its population increasing very fast. The dryland regions increase stamina in the people resulting in large sized families. Average family size in Rajasthan is 6 persons which vary in different regions of the districts. District is administrative area where different climatic zones may exist. Land use pattern is greatly affected by drought. Clearance of vegetated area for meeting agriculture requirement remained continuous feather since man started settled life. Cutting of trees for various human requirements is also continuous phenomenon and this process has reached to saturation where most of the land has been denuded for human requirements and now this is last stage of environment degradation, which needs immediate reversal. In cultivation practices severe degraded lands became barren and waste lands or large scale deforestation brought this condition. Even forest lands are only demarcated areas where forest trees are rare. Thus land use pattern has deteriorated the complete vegetation system and atmospheric features are free to play their devastating role.

Land is natural resource which is limited and cannot be increased with human efforts. Putting land under productive use is the necessity for survival of human beings on the earth. Large scale deforestation has extinct wild life which was main source of human requirements but now land productivity can only help the people to survive on earth for which productivity of the land need to be retained in any manner. Forests are necessary in 33 percent of area to maintain the environment and ecology of earth surface as natural factors become positive for human requirements. Therefore, cultivated area need to be retained for meeting the human requirements which is only possible when forest area of requisite extent are revived. Droughts are disastrous for cultivated lands and denuded forest areas are also dangerous for soil erosion through wind and rains. Atmospheric extremes are the impact of loss of vegetation from earth surface. Human disastrous acts are supplemented by natural factors in destruction and droughts further exaggerate the land degradation.

Present land use pattern of Rajasthan is not favourable for sustainable use of land for all requisite purposes. The present rate of land degradation of all the categories of land is creating circumstances for extinct of human beings from earth surface and its reversal is necessary. With present rate of land degradation total land of the state is feared to be vanished for any vegetative use in five decades at the most. The severity of droughts became uncontrollable with land degradation as per present situation. Presently 60 percent of total geographical area is cultivable but half of the land has been largely degraded to the extent where cultivation practices are mostly uneconomic as input cost is not possible to be yielded. In this situation total cultivable land need to be reclaimed completely to feed state population increasing at very fast rate. If land degradation continues further, it is not possible to imagine the condition of meeting human food requirements.

The present land degradation is treated as the impact of drought but it is not completely the correct narration. Droughts disturb the fertility of land but its severity increases manifold when human interface allows to land degradation as consequential impact of drought. Mostly single cropped areas are most vulnerable to decrease of soil moisture as land remains without any obstacle and natural factors play their destructive role On the contrary double cropped areas remain without cover for the limited duration of 3-4 months. In this period, cultivators plough their lands to make further acceptable to crops as harmful insects extinct in hot temperature. In case of single cropped area, natural factors get more time for land degradation which further aggravated during the summer. This is the main factor of moisture loss and serious land degradation.

The present land use pattern is ideal for reclamation as cultivated and fallow land is about 60 percent of the geographical area. Similarly forest land, pasture land, barren and waste lands account for more than 30 percent of state geographical area. Remaining 10 percent area is not available for any productive use. Cultivated area and other area are inter-dependent on each other as cultivated lands can be available for sustainable cultivation practices only when the forest cover over more than 30 percent of land area is reclaimed under forest and vegetative cover. Cultivation practices are only possible when forest and plantation area is undamaged and tree cutting is done only when it is dead. Live tree cutting need to be checked for management of ecology and environment of the state. Plantation area needs to be covered with dense forest and old dead trees need to be filled with new sapples. If plantation area is fully equipped, the cultivated area would also revive with changed position and adverse impact of drought would be within control.

5.7 IMPACT OF DROUGHTS ON CROPPING PATTERN

Cropping pattern is governed by rainfall specially the fain fed crops. Post monsoon crops are also possible when rainfall remained normal or with some deviation. Cultivators plough their lands in the hope of good rains and their hope remains active before rainfall. In case of widespread rainfall, cultivators sow various crop species according land suitability. If monsoon is delayed, the cultivators prefer to sow short duration crops where yield is low in money terms but cultivators are able to manage their livelihood. Monsoon uncertainty reflects the cropping pattern in many ways. In case of widespread rains, the sowing is done in maximum possible area and further spells are delayed or deferred, crop failure is certain. Relationship of crop and rainfall is unique and germination and growth of crops is dependent on regular spells of rain. If withdrawal of rainfall is before time, the crop damage is eminent but fodder availability is certain.

Cultivators of Rajasthan are well versed with the erratic nature of monsoon and try to avail maximum from every short of events. In case of widespread rain and delay in second spell crop failure is certain and in such condition cultivators try to sow for the second time by quick growing crop species to get something for domestic requirements. Cultivators try their level best to accommodate with monsoon events and if there is widespread rains in terminal duration, cultivators try to avail the facility to grow crops which may survive with moisture retained. The cultivators have to correlate with the monsoon to the possible extent and get agriculture produce of their own requirement and sell surplus produce for others. The cultivators of the state though mostly illiterate but understand to accommodate with the nature to the possible extent. This is the way to grow crops in the state facing all the adversities.

Central Arid Zone Research Institute, Jodhpur has developed various crop species, which can survive in dry spells of monsoon even for fifteen days and such efforts have helped cultivators to great extent. Agriculture university and their research laboratories have also developed various crop varieties which can be successfully sown in short duration. These drought resistant varieties have helped the cultivators to face the vagaries of monsoon to vital extent. Cropping in various uncertainties could be possible with the help of various technological devices. Adjustment with nature is the necessity of Rajasthan cultivators and this is the condition where crops are grown to feed the population.

Jaisalmer district is known for lowest rainfall in the state. The cultivators of this district are cultivating post monsoon crop species with moisture conservation through khadin system. Khadins are formed in the vicinity of hard surface structures like hills, hillocks, mountain ranges, pavements, pediments and even sand dunes of more than 9 metre height. Run off water from such structures is carried to sloppy fields where moisture is retained during monsoon session. After withdrawal of monsoon rabi crop species like wheat, gram, rape and mustard are successfully grown in the fields. Cultivators even use such khadins for late kharif crops varieties of short duration Drought effect over such lands is mild and cultivators get food grain slightly lower than average yield of irrigated lands.

Now khadins are being prepared with technology support where yield rate is equal to irrigated lands. This area is not treated irrigated area but moisture cultivation is successfully carried out. Some areas of Jaisalmer district are linked with canal irrigation system where all variety of crops can be grown. Droughts are not any new event of Rajasthan and people try to settle down cropping pattern as per monsoon conditions. This is not possible for human beings to change natural factors but they can very well adjust their activities as per rainfall conditions and that is the only way to survive in most critical environment where droughts are recurrent phenomena dwindling the human efforts.

6

Impact of Droughts on Environment

Droughts are severe natural disaster, which affect environment and ecology of the state. The environment degradation in Rajasthan from droughts is quite significant in view of its constant occurrence and there is no time for replenishment. One drought with widespread impact usually takes about five years to coup up the environment but recurrent droughts have most disastrous impact on the environment. Droughts are not only factor for environment degradation but human interface is also major factor for magnitude and impact of drought to very serious consequences. This is the reason for stagnant of state economy. Environment maintenance is necessary for development of flora and fauna over the earth surface but the present level of degradation has reached to most critical stage

Deforestation is most critical factor of environment degradation, where uncontrolled felling of trees has created most devastating act of environment degradation. The intensity of droughts has been increased manifold as soil erosion from water and wind created major problem in arable and non arable areas. Wasteland and barren lands have further been deteriorated where plantation activities are not possible. Even grown trees are cut for human requirement of fuel wood, fodder and timber. In cultivable lands fallow lands remained unused for years have become wastelands which are not possible to use for crop growing. The cultivated lands are losing their productivity in view of salinity and alkalinity. All these destructive acts of human interface have degraded environment to vital extent.

With the loss of vegetative cover, extremes of temperatures are being recognized by people as hindrance for vegetative growth and development. Extreme temperatures have helped in evaporation of moisture from earth surface and sub-surface helpful for vegetation. Trees were helping in wind erosion and movement of sand dunes from one place to another. After large scale deforestation, the earth surface have become denuded and fertile soil is also disturbed with wind storms.

Dryness is prevailing in most parts of the state which is harmful for human, livestock and vegetation. Every drought further intensifies these problems to great extent. Humidity presence in environment is decreasing constantly where vegetation germination and growth is seriously obstructed the developmental process.

About two-third part of the state is covered under desert development programme and drought prone area programmes, but no significant results could be achieved. Germination and growth of tree plants is quite difficult in degraded environment. These programmes have been introduced in the state as national programme of drought proofing. Uncontrolled grazing of livestock in plantation areas are critical to combating drought prone areas. Similarly water harvesting activities through watershed process and anicuts in most parts of the state are constructed for increase in ground water level of the area and in some parts where land surface is hard, irrigation is also possible, but these efforts could not yield much. During drought years, rainwater is not sufficient to increase water level and water draft from these sources depletes water level. In most parts of the state, water depletion has reached to most critical level where further draft of water would extinct these resources.

Rainfall has become most erratic and uncertain which has further intensified cultivation problems. This is the resultant impact of large scale deforestation, which helped clouds to attract over the state regions. Deforestation has devastated the environment system of western Rajasthan where wind storms carry dust in the sky quite high and stay as barriers for clouds. Deforestation has created various environmental problems in the state as rainfall has been restricted to few showers but sometimes create havoc through floods. Soil strata of most of the state is loose and no drainage systems exist in most parts of the state. Most of the regulated drainage systems have been blocked by mobile sand dunes.

Erratic nature of rainfall has created uncertainty in cultivation practices and there are few no-drought years when total state is free from drought disaster. Erratic nature of rainfall is responsible for recurrent droughts, which are unable to predict. Rainfall failure has multi-faceted factors from delayed monsoon, failure in the middle of the rainy season when it is necessary for germination and growth of plants, uncertainty in further spells of rainfall and early withdrawal of monsoon. Almost 80 percent of the cultivable area is monsoon fed for cropping but in case of drought conditions, cultivators remain main sufferers. There are no other income generation activities and cultivators of single cropped area are completely dependent on monsoon. In view of severe environment degradation, even little deficiency in rainfall is resulted in crop failure.

Thus human interface and natural factors disturb the environment to great extent. Even the erratic nature of natural factors is also the impact of human interface to great extent as all the environmental factors have been deteriorated to vital extent. Environment degradation takes long time when the human interface disturbs the ecological balance of the region and natural factors start their negative role. In the context of Rajasthan state, people could not resist destructive acts and government remained spectator to the uncontrolled human acts. Droughts have further intensified the problem.

6.1 PREDICTION MEASURES OF RAINFALL AND DROUGHTS

In normal conditions, occurrence of rainfall starts from last week of June to first week of July and continues functional up to mid September to end of the month. Post monsoon rains do occur from October to March as per atmospheric conditions. Usually there are 10 to 22 rainfall days in the state varying from western part to southern and eastern parts. 61 percent of the state geographical area is desert where rains are quite meagre in comparison to remaining areas. However drought conditions prevail through out the state irrespective of rainfall conditions. The geo-physical conditions of the state are largely varied as Jaisalmer district with 18.55 cm rainfall with favourable spells can yield good crops and Banswara district with more than 90 cm rainfall can have similar results in normal rainfall conditions.

Requirement of rainfall for agriculture, hydrology and meteorological factors remains varied as cultivation requirements need timely rains with regular spells, while hydrological and meteorological factors require normal rainfall in the total monsoon season irrespective of its duration and intensity. Accordingly agricultural drought is the resultant factors of disturbance of monsoon system from beginning to terminal spells with irregular intervals. Agriculture drought is most severe as this affects large section of population and related to their livelihood. Hydrological and meteorological drought have comparatively lesser disaster in comparison to agricultural drought, but when all the factors combine together, the severity of drought is most critical in the affected area.

6.1.1 Prediction measures of Rainfall

Prediction of rainfall originates from observations in the Indian Ocean as this is the most scientific system of the prediction but this prediction relates to country as a whole. Such predictions are possible in the month of April. The predictions are based on the atmospheric conditions over the ocean and land areas where wind waves move frequently from ocean to land and vice versa. These predictions are most crude in terms of occurrence of rains in the country in the monsoon duration. The deviations are drawn from the normal rainfall trends in the country. These predictions vary very little as country is very vast and occurrence of rainfall cover the land area of the country in phased manner.

Prediction of rainfall for Rajasthan as a whole is also done in the same duration but these predictions remain most crude in terms of arrival of monsoon in the state as per its trends and occurrence of rainfall in relation to normal rainfall conditions. Further predictions are made when the monsoon struck the land area in Andaman and Nicobar Islands and later in Kerala state in first week of May. Next round of predictions are made on the basis of arrival of monsoon in Mumbai from Arabian Sea and from Bay of Bengal. With the help of satellite maps the forecasts of rains are made for country as a whole with state wise forecasts. When monsoon clouds move with the wind directions, the predictions are made on the basis of the direction of wind waves active at the time of satellite observations. Appearance of clouds over the country as viewed from map helps people to understand the trend of movement and expected rainfall in the given duration.

The monsoon predictions are analyzed on the basis of wind trends and atmospheric pressure. Extremity of hot temperature create vacuum in the area and predictions are based on such conditions. Such predictions vary significantly as the wind directions keep on changing with atmospheric features and there are sudden changes to blow clouds in other unexpected directions. Rajasthan is largest area of the country spread over 10.41 percent of country area. In addition, Aravalli ranges divides the state in two different climatic zones, where western part remains most deficient in rainfall. While covering long distance up to Rajasthan, the clouds are mostly empty and the state is able to receive mild rains as last spell or concluding part of their passage.

Rajasthan state gets rains from both the systems through Arabian Sea and Bay of Bengal but the state is situated in the terminal range of monsoon clouds where the pressure of clouds is decreased significantly. Thus the earlier predictions are generally cursory for Rajasthan state while accuracy of daily forecasts remain more authentic but such predictions are relevant for the state as a whole or some regions which vary greatly. Cultivators start ploughing their fields in the month of June and complete total process as per requirement before occurrence of rains. The earlier predictions are very little relevant in terms of their authenticity. The cultivator community is also well aware about the relevance of such predictions and keep their land ready for sowing after first rains.

The cultivators of the state carry out their sowing process after first spell and select crop varieties in the basis of its occurrence trends. The predictions trends on daily basis are relevant and these are near authentic in terms of area coverage. Actual situation of rainfall depends at micro level in terms of its intensity and coverage. If first spell of monsoon rains is widespread covering total state area, cultivators cover larger area under cultivation otherwise restrict their task to high yielding areas as per their judgement Since sowing process starts after witnessing its occurrence, cultivators take quick decision keeping in view the intensity and magnitude of the area coverage, which is relevant.

Cultivation activity in Rajasthan state is most critical and cultivators are the only judges of their acts and further consequences of their acts are to be borne by them. The earlier predictions of rainfall are most tentative and cultivators do not give much cognigence to such predictions but watch the occurrence of rains quite patiently. In case the first spell of monsoon is delayed, low or erratic, the sowing practices are deferred till next spell and this sort of disturbance is the symptom of monsoon failure. If next spell of few days delay is widespread, the sowing practices are continued as usual. There are many obstacles in the cultivation practices as loss of inputs is borne by the cultivator.

6.1.2 Prediction of Drought

The prediction of drought remain disguised till the occurrence of first spell of monsoon rains, though rainfall forecasts are available well in advance. The predictions of rainfall and droughts are generalized and cannot be treated its impact over any specific area or region. Even the daily rainfall forecasts cover larger area and micro level prediction are never authentic. The monsoon is governed by so many atmospheric factors and localized predictions cannot be authentic in this regards. Even the cultivators do not rely over such predictions or forecasts and watch occurrence of rains in their own locality. Cultivators usually consider everything individually while consulting their co-cultivators of their village but beyond their village boundary, no such consultations or observations are made.

Problem with monsoon is its inability to predict and their predictions are hardly accurate. This is the reason that crop failure is severe in terms of area spread and intensity. The special feature of monsoon in the state of Rajasthan is its arrival in the state in time or with some delay but first rain is usually widespread. If monsoon is delayed by a fortnight or so, the cultivators opt to sow in lesser area by avoiding less productive lands. This is the reason of variation in net cropped area, which varies on yearly basis. The spells of monsoon from its occurrence to conclusion remain disguised in total monsoon season and earlier predictions are never authentic at local level. The disturbance in first spell is resulted in low cropped area but sowing is must for cultivators and everything is uncertain from beginning to end and cultivators rely over their own observations from occurrence of rains in their own village.

Usually their four to five spells of rainfall and crops germination and growth is also correlated with these spells as crop can survive even for a fortnight delayed from its regular trend. With this rainfall behaviour, cultivators continue their agricultural practices. In addition it is the helplessness of the cultivators to sow crop species even in worst years as monsoon trends are never behave as per predictions, estimations and forecasts. Sometimes, the rainfall erratic in the beginning becomes regular in further spells and cultivators get substantial return. Thus total cultivation practices are continued in the circumstances of uncertainties and usually cultivators are benefited with their decisions and preparedness.

Cultivation practices are most critical in view of erratic behaviour of monsoon and its failure can collapse the cultivator community as per its spread. Droughts are predictable to little extent but its local impact is known on failure of monsoon and crop damage. Usually such situation arises in most part of the state in a cycle of five years but unable to predict in advance. Analysis of past trends of cultivation practices are evident that crop sustain in low rainfall if its spells are timely and regular. The situation in totality is possible to be assessed when the crop is harvested.

The state government issues forecasts of cropped area and production from sowing to harvesting levels, which is short of vigil of cultivation practices on constant basis. The first forecast is about the area sown, which reveals that the rainfall trends in total state, and area sown in each districts is reviewed in terms of coverage. These figures are further analyzed in terms of earlier reference and trend of sowing in evident of good or bad rainfall and cultivator's acceptability for sowing. Usually there are three to five forecasts for different crops, which reveal trend increase or decrease in production in the state.

At the state level area sown is response of cultivators which decides future course of expectations. In case the cropped area is quite low, the production and productivity is bound to be in the proportionate manner and when the sown area surpass the five year past trend, the response of cultivators is treated as overwhelming. Evan the impact of drought is known in the month of September from the crop damage position, which is ultimate position of state as a whole and impact of drought is known in district specific terms. Even the impacts remain varied in a district and such trends are further correlated with the intensity of damage, area affected, villages affected, population and livestock affected. Even the estimated crop damage is known at the state level for further action in view of area affected and crop damage.

In case drought prediction are known to accurate level, the loss of cultivators can be saved to great extent, but the degree of accuracy is not that much reliable and even cultivators make their own assessment as local conditions and take appropriate action. Thus the total cultivation practices are still carried out on conventional pattern as prediction of rainfall and drought conditions are not possible to accurate level.

6.2 INDIGENOUS KNOWLEDGE OF WEATHER FORECASTING

In rural areas the electronic and print media has made total connectivity and whether forecasts and predictions are made from time to time to help cultivators to the most possible extent, but still cultivators prefer their own assessment of rainfall and occurrence of drought. The cultivators are generally less literate but well versed with the cultivation practices. They believe in demonstration and replication methods to large extent. Various crop varieties suggested by the experts are accepted by cultivators by viewing closely their acceptability in the soil and environment. Introduction of various varieties of oilseeds, food grains and pulses. Even commercial crops are being carried out by cultivators after witnessing the results in similar conditions.

Cultivators are best judge of their land and know the level of degradation and productivity level can be attained in positive impact of all natural factors. They sow crop varieties while viewing the impact of yield and their capacity to bear the climatic fluctuations. The technological developments have helped the cultivators in general observations of rainfall and its occurrences based on predictions for preparedness of cultivators for planning their cultivation practices. However, the predictions are usually deviated to significant extent. In case of good rainfall conditions, drought prevail over larger area of the state and crop failure is significant. The problems with the cultivators is to spend money on inputs and take reliable varieties of seeds, pesticides, fertilizers etc at their own. In case of crop failure, the loss is irreversible as their labour is futile and input loss is the additional loss. In case of crop damage less than 50 percent, cultivators are supposed to bear the consequences.

Cultivation practices in the state are most risky and less remunerative. If wage component of cultivator family is calculated with prevailing minimum wage rate, they are generally looser even in normal year. The risk factor is quite severe and cultivator community never removes out of the clutches of the natural factors. In case of severe drought conditions, the loss to cultivator is unaccountable as government at its best provides wage employment for the duration of next monsoon if severity of drought is most severe. Loss to human beings, livestock, land degradation etc. are to be born by themselves. In this miserable conditions, cultivators continue their traditional activities as they have no alternative sources of livelihood.

Cultivators have their own prediction system based on natural factors and they rely those most. There are various indigenous modes of weather forecasts and usually rely over them in view of their reliability. These predictions are usually based on lunar conditions, which they view closely in April to June specially on full moon night. If there is yellowish circle around the moon, the year is treated as good rainfall year. Lunar impacts are also studied by meteorologists for assessment of tides, which rise maximum on full moon night. The predictions based on lunar impacts are near accurate and cultivators usually make their own forecasts. Lunar impact has widespread belief amongst the rural people as such impacts are also studied by astrologers.

Direction of wind is another mode of prediction of rainfall conditions based on indigenous belief. These predictions are possible within a fortnight of rainfall and this system is part of meteorological studies as winds directions from sea to earth and vice versa are relevant to monsoon occurrence of changed directions. This indigenous system is authentic and reliable but scientists have shifted to satellite system never consider this trend. Winds are helpful in carrying monsoon clouds or may divert to other directions due to atmospheric fluctuations. The indigenous knowledge is usually simple version of scientific systems and cultivators rely over them after their own assessment about the predictions from their local approach.

Other indigenous knowledge of monsoon and drought prediction modes are many in different parts of the state. Cultivators rely over any knowledge after testing its viability and few years of accurate results develop their faith in the indigenous knowledge. There are various modes of rainfall predictions on very day or week. These modes are based on animal behaviour, actions of birds, ants, etc. The problem of cultivators in this sphere is multifaceted as they want some reliability of predictions. The meteorological predictions are rarely accurate in view of their constraints as atmospheric pressure change the situation that is unable to be assessed.

The indigenous knowledge is combination of traditional system, already tested by their ancestors and there is no reason to disbelieve over them if they are accurate to some extent. The meteorological science is not that much authentic as physics, which are proven in all the conditions. The meteorological science have various limitations and even the satellite information based predictions do not happen in predicted manner. In such conditions, cultivators and other rural people rely over their indigenous knowledge which is largely prove as predicted. However, the indigenous knowledge is quite varied from area to area as people of each region have their own mode of prediction.

The indigenous systems of prediction are not merely the myth or ghost stories which the people believe in view of fear or mishappenings. Traditional knowledge is accepted by people in view of their reliability and trust. The drought and monsoon failure are very unfavourable factors for cultivators and rural people and any knowledge is proven with reality is welcomed by the people. Cultivators want some accuracy and exactness, which helps them in taking decisions well in advance. The cultivators have been assisted by agronomists and research institutions to take appropriate decisions about the crop varieties adaptable to drought conditions.

The cultivators pay attention to meteorological predictions and also consider their indigenous knowledge and analyze both types of predictions. With this approach they are able to get some clue to reach at the appropriate stage but take suitable decisions after occurrence of first rainfall. With this approach the risk factor is minimized to some extent. The sowing activities start after first rain and the predictions are also there to take suitable decisions about the crop varieties to be sown and also the area coverage. Usually cultivators with meagre operational holdings have little options for cultivating their fields or remain fallow. Cultivators take decision on the basis of weather conditions in their own village and also in the surrounding villages. This knowledge is sufficient for them to take suitable decision about the future of monsoon in their region in general and their own village in particular.

Further spells of monsoon are unable to be predicted at local levels and all the predictions are broadcasted for larger area or regions. Therefore future consequences of monsoon are unable to be predicted through any source. This is the reason of large scale crop damage in case of most severe drought in any area or region. This situation itself is evident that predictions for total monsoon duration and accuracy in rainfall spells is unable to be predicted. The atmospheric changes are most varied with frequent changes which the experts of meteorological department. From cultivators angle, they cannot postpone seed growing on the rainfall predictions in view of lesser reliability. This is the reason for large scale crop damage in the state.

The cultivators of the state are quite courageous to sow various crop varieties over-looking the predictions. Now he cropping pattern, technological development and demonstration impact of success stories. The cultivators have little fear of crop damage and remain ready to face drought for any stage from beginning to crop harvesting. However the cultivators remain feared about monsoon failure and felt sigh of relief when they are able to harvest their crop carrying their agricultural produce at home. The degree of uncertainty about the cultivation practices is close associated with the cultivation practices. There are instances when the standing crop successfully cleared all the fearful stages but rains do occur in pre-harvesting duration damaging total crop of the village. Therefore predictions for monsoon and drought have various faces from beginning to the end.

6.3 SATELLITE INFORMATION SYSTEM AND PREDICTION METHODS

Satellite information system is technological improvement in the field of monsoon prediction and through these predictions it is possible to forecast drought occurrence in case of failure of monsoon. Satellite move around the earth over the Indian sub-continent and take photo of the atmospheric conditions prevailing over the country and surrounding oceans. This is latest technological device which placed India in to prestigious position among the world community. Daily photo snaps from the satellite moving over the sub-continent once in a day are taken by the earth station and analysis thereof is the source of weather forecast for next twenty four hours and trends of whether for coming duration. These photo imageries are broadcasted over the television along with low pressure zones in the ocean and surrounding areas. The predictions are done by the experts on the basis of wind directions and low pressure zones, which help in making assessment of the intensity of the ocean developments.

This system has one limitations that photo imageries taken once in a day are the location of the some event on land or the surrounding ocean and seas. The predictions made on the basis remain based on the events at that particular time but sudden changes in atmospheric conditions negate the predictions to vital extent in terms of the impact area. The atmospheric changes are sometimes very vital, which cannot be predicted. Low pressure zone developed in the ocean or surrounding seas gets any direction at the time when the satellite is getting the photo imageries. The analysis is made on the basis of the wind directions and predictions are made about the occurrence of events in the said directions. This is the system of analysis and has its validity which usually happens.

The changes which take place in the later duration are related the decrease or increase of the intensity and the wind directions. In case of development of any cyclone in the ocean the intensity and direction changes suddenly which falsifies the predictions made on the basis of the photo imageries. Sudden changes in the pressure decide the intensity and atmospheric changes are capable to shift the wind direction. Thus the forecasts made on the basis of analysis of photo imageries change to vital extent. Though the changes in the pressure or wind direction are bound to happen, which develop suddenly due to atmospheric conditions, but the events do occur irrespective of intensity or location.

The situation of changes in the predictions are the peculiarities of the meteorological system and if these limitations are well known in advance, there is no problem in understanding the limitations of this scientific device. In most developed countries use of multiple satellites is the mode of accuracy of the predictions which details the sudden changes and alert the people of he area to remain cautious of the event. Thus their predictions are further supplemented through revised and latest forecasts of the events. The Government of India is well aware of this problem and trying to reach at the accuracy level to gain faith of the people.

Usually rainfall forecasts of day change for some identified region and the changes made due to changed atmospheric conditions lose the faith of the affected population. The predictions are made on the basis of photo imageries from satellite, have some lacunae as experts are new to the system. There is no fault in the satellite photo imageries as depictions of events are made on the basis of actual behaviour of ocean and atmospheric events. The interpretation method of these events need to be assessed on the basis of ocean and atmospheric conditions. The earth station receives the information from the satellite has also no problem as the system has no defect but the human error is the problem.

The scientists engaged in analysis of the events need training and their capacity building is essential to make actual forecasts of the events. The accuracy is the most critical factor as the information dissemination is relevant for country as a whole and error in analysis diminishes the reliability of the system. The predictions need to be done by senior scientists for making the relevance and gaining faith of the people. The cultivation practices in Rajasthan are slightly changed from other regions as sowing is done after rainfall. If the sowing is carried on the basis of predictions, the cultivators will remain loser and their faith with the system would lost, which is the matter of concern for the scientists. The interpretation system need to be revitalized and defects in the analysis system need to be assessed in terms of their process and analysis mode.

The satellite system and earth station is most modern to the meteorological science where role of scientists is most critical. It is further valuable that the information is useful for most of the people engaged in cultivation practices. The cultivation practices in India are still based on conventional system and real help of the scientist community is to gain their faith by broadcasting information at most accurate level. If this system is deviated from its principal role, the cultivators are bound to revert back to their indigenous knowledge. The cultivators are facing most hardships as their total cultivation practices in Rajasthan are based on uncertainties and if the scientists are unable to perform their role in most professional manner, expected from them, there is no necessity to incur huge expenditure. Thus the role of scientists is most vital and they need to prove their worthiness.

This is bare truth that nature is bound to play its role as usual and rainfall will occur at the places where atmospheric conditions are favourable. The prediction of monsoon are helpful for the cultivators to keep their preparedness according conditions. The reliability of prediction system is helpful to keep advance preparedness to save damages. Cultivators are supposed to carry out their agricultural practices whether there are monsoon predictions or not. Cultivation practices were being performed in pre-satellite era also and will continue in future as this is necessary. The prediction system is helpful to cultivators to have advance preparations for facing the challenges of the rainfall behaviour. The accuracy of the system is most critical for its reliability. If the meteorological scientists are unable to prove their worthiness for the specialized knowledge, which they claim is the main issue and its reliability is at stake.

Mostly cultivator community do not rely over the monsoon predictions of meteorological department in view of the fact that they are not reliable to legitimate extent. Cultivators were earlier dependent over their indigenous knowledge and still compelled to base it in their own interest. The role of scientific community for their predictions in pre-monsoon duration and on ensuing days need to be helpful to cultivators as they can carryout their advance practices to gain maximum or minimize their losses to possible extent. The cultivators will rely over the system if it is true to vital extent. In case their predictions are not true to such extent, how cultivators can accept their suggestive practices if there is no surety of any extent.

The prediction system with the help of satellite imageries need to be reviewed by senior scientists and find out their defects in the prediction pattern. The short comings remained in the interpretation method need to be analyzed thoroughly to find out the reasons of mistakes in their observations. The scientists are not immune in the country and they are supposed to arrive at the level of accuracy where their predictions are proved true to legitimate extent. If the deficiencies are removed in future predictions, the faith will be revamped and cultivators would have no hitch to rely over the system. The scientists if unable to perform their role in successful manner, their position is always questionable.

The cultivators may rely over the most authentic scientific system is possible when the scientists are able to prove their existence. The help they expect by cultivators in terms of their reliability and ability to minimize their losses is only possible, when these are proved true as cultivators accept anything to be valuable in their task. The sufferings of cultivators are most critical where their total practices remain at stake. Real help to them is to release true picture of events and create the environment, where cultivators could gain their faith. The system of prediction need to be reviewed in details and make every effort to prove their existence in professional manner.

6.4 RELIABILITY OF WEATHER FORECASTING AND DROUGHTS

Drought is one of the severe disasters, which occurs due to monsoon failure. Drought is able to be assessed in advance as here are multi-faceted factors responsible for monsoon failure from its occurrence to retreat. Cultivators realize the drought conditions visible over their crops. Whether forecasts are done before two months of occurrence of rainfall and this process continues till its retreat. Atmospheric conditions play their role by occurrence of pre-monsoon rains which are also possible to be forecast before twenty four hours. The atmospheric conditions also play their role after retreat of monsoon. Whether forecasts are usually possible with the behaviour of atmospheric conditions on the ocean and land surface. These forecasts are based on low pressure zone formation over the land surface with extremity of temperature where winds blow up due to extreme heat and vacuum is created over the land surface. This vacuum is bound to be filled by cold waves which move to vacuum area very fast.

Simultaneous system occurs in the oceans where the extremity in temperature form water vapours which blow high with the wind support and move towards land surface. Sometimes the winds form very severe low pressure zone and rainfall occurs in the sea or ocean without moving at far distant places towards land surface. Weathers are formed with atmospheric changes and sun plays vital role in view of its location. Summer and winters are weather conditions depend on the location of the sun over the earth surface. During summer, there are vary significant changes in the behaviour of the oceans and seas and sea water raises high in the form of vapours. The low pressure zones are formed over the sea and earth surface which are causatic impact for wind waves blowing from ocean to land and similarly from land to oceans.

With extremity of heat of the sun clouds are formed which move towards land surface and rainfall occurs. This is natural system and occurrence of rainfall is the impact of these atmospheric conditions. Rainfall occurs over the land surface as per wind directions, which is also well designed to atmospheric conditions. In Rajasthan rainfall occurs through wind waves reaching over the state through Arabian Sea and the Bay of Bengal. Variation of rainfall is the impact of Aravalli system, which divides the state in two different climatic zones. The rainfall remains erratic and uncertain in total state and droughts do occur due to failure of monsoon.

Forecasts of weather are the necessity of the state for knowing its behaviour in total monsoon session and its occurrence in different climatic zones of the state. The Meteorological Department is responsible for prediction of monsoon over the country as a whole and states which includes Rajasthan also. The forecasts are based on the behaviour of atmospheric conditions and these forecasts help in cultivation practices as significant cultivated area is cultivated through monsoon rains where single crop is possible. The cultivators having unirrigated cultivated lands depend solely on the occurrence of rain and failure of rainfall has severe consequences for livelihood. About 80 percent of cultivators are dependent on monsoon based cultivation practices and occurrence of rainfall in time with assured spells has meaning for them. In case of good rainfall about 90 percent of cultivable area is sown with adaptable crop varieties.

In case of monsoon failure the area sown is decreased to less than 50 percent depending on the intensity of failure over the state. If monsoon failure is vital covering total state, the drought conditions prevail throughout the state and there is every possibility of famine conditions if relief measure are not initiated immediately. Rainfall forecasts help cultivators by leaving their lands fallow in severe drought conditions to save input loss, which is also vital for poor cultivators.

Weather forecast help cultivators in changing cultivation pattern by sowing short duration crops or avoid monsoon crops and await monsoon behaviour where they can grow post monsoon crops if sufficient moisture is possible.

Weather forecasts are helpful to cultivators and also becoming their necessity in view of erratic and uncertain behaviour of monsoon. The problem with most of the cultivators is their dependence solely over the land as this is the only source of their livelihood. The necessity of the cultivator community is the accuracy in weather forecast system to help them take appropriate decision. Cultivation of monsoon crops is most critical as failure of monsoon has various stages and simply the weather forecast of occurrence of rainfall in total monsoon session has no relevance to cultivators. Chances of crop failure remains active from occurrence of first rains to its withdrawal and every spell of rainfall has relevance to cultivators.

The problem with weather forecast is unreliability and it is most critical factor for the sustenance of cultivation practices in the state particularly monsoon crops. Post monsoon crops are based on monsoon behaviour as cultivators are able to assess the situation of moisture in their fields, availability of ground water resources in their access and canal irrigated regions. Problem in monsoon crop is its dependence on the occurrence of rains in most part of the state and weather forecast need reliability in terms of its occurrence and regular spells. In case of widespread rains of first spell, cultivators cover larger area in the hope that marginal and degraded regions would supplement their productivity to vital extent and they try to sow crops in total land cover their operational holdings. In case of deviations in further rainfall spells, the crop failure becomes inevitable and loss is severe.

If weather forecasts are accurate, cultivators are most benefited for saving from crop damage. Cropping require inputs and labour of cultivators and in case of total damage, cultivators are helpless to manage their livelihood as total annual requirement of family is associated with rain fed crops. The environment and ecology of the state has been deteriorated to significant extent and erratic rainfall has further aggravated the problem. In this background weather forecast is vital for the cultivators. Weather forecast is possible with the help of advanced technology developed in the country but its reliability is the problem. Weather forecasts for occurrence of rainfall and prevalence of droughts is still unable to be predicted with degree of accuracy.

Prediction of droughts is based on occurrence of rainfall and failure of rainfall is closely associated with the drought conditions. Intensity and magnitude of rainfall and of drought is most relevant for the cultivators and if forecast is possible before occurrence of rainfall, preparedness for facing drought is possible for the cultivators and other affected people of the state who are the ultimate sufferers of such mishappenings. Analysis of magnitude and intensity of droughts during 1997-2003 reveals that during the year 2002-03 all the 32 districts of the state were severely affected with drought. During 2000-01 also all the 31 districts were gripped with the severity of drought as Karauli district was established in 2002. During 1999-2000, 26 districts were affected by drought and in 1997-98 the number of drought affected districts was 24. In 1998-99 and 2001-02 the drought affected districts were 20 and 18 respectively. The intensity and spread of drought can very well understood with the past trends.

Forecasts of rainfall and drought is not possible to full accuracy and this is the main constraint. Cultivators grow their crops in the hops of good yield but crop damage remains the main problem and state government is also unable to take advance action to face drought conditions. There are various most severe drought prone districts which suffer from drought problem on regular basis as environmental conditions are mostly deteriorated and even moderate deviation in rainfall creates severe problems in these districts. The crop damage is vital in case of failure of monsoon which usually fail in the last few spells.

6.5 VALIDITY OF YEARS OF PREDICTION

Annual rainfall predictions made show 5 to 10 percent rainfall deficiency. Such predictions are relevant for country as a whole but in case of Rajasthan, the predictions made for annual duration do not prove their validity. Year 2002-03 was the most critical year from drought point of view when all the 32 districts were seriously affected with the most disastrous drought. In this year the expenditure incurred on drought relief measures was more than the total expenditure incurred on drought relief from 1951 to 2002. Even the annual predictions predicted 10 percent deficient rainfall, whereas the state average rainfall was 56 percent below the normal rainfall. The district specific variations were up to 76 percent below the normal rainfall. This is the main constraint with the prediction system where accuracy remains doubtful to vital extent.

The validity of annual prediction system is very critical where the reliability of forecasts remain very far from the actual situation. prediction of drought is not possible with existing system as the impact of failure of rainfall is known quite late from the actual occurrence. At that stage the actual damage has been happened and no retrieval is possible in terms of loss. Actually, the situation of Rajasthan state is quite critical from the point of view of occurrence of rains, where country wide predictions have little relevance. It is certainly true that deficiency of rainfall over the country remain between 5 to 10 percent and that is regular feature of the monsoon, which can deviate in this range on both the sides. In this situation, it is necessary to consider whether the prediction for Rajasthan is possible with district specific details. Usually the present system of prediction is not feasible to this extent.

Rajasthan has typical climatic conditions where the occurrence of rainfall remains varied in terms of quantity, spread and time and all these factors are vital for the state. The drought has become regular feature of the state and if prediction of annual occurrence of rainfall is not possible for various climatic zones of the state. It is quite difficult to face with the drought, which has dwindled the economy of the people of rural areas as well as the state. The state domestic product is seriously affected with the drought disaster and there seems no way even to predict the occurrence of drought. The variations in rainfall are most critical for the state and there seems no way out to overcome the situation with the existing system of forecasts and predictions. If the most developed technological devices are not able to predict the rainfall and drought conditions, it is misfortune of the system, which remains futile for the conditions of Rajasthan.

7

Measures for Declaration of Drought Affected Areas and Relief Measures

The drought affects the state regions through crop damage, shortage of drinking water and disappearance of surface water harvested in water bodies. The drought is regarded as an extreme form of agricultural drought, resulting from meteorological and hydrological droughts where food, fodder and water shortages are so severe that large number of people become unhealthy or die. Famine disasters usually have complex causes often including war and conflict. Although scarcity of food is the main factor in a famine, death can result from other complicating influences such as diseases or lack of access to water and other essential services.

Socio-economic impact of drought correlates the supply and demand of goods and services with the multifaceted effect of food and water shortage. Shortage of essential commodities and services like water, hay, food and electricity is weather dependent and shortage is often happen or may be created. The concept of socio-economic drought recognizes the relationship between drought and human activities. Poor land use practices exacerbate the impacts and vulnerability to further drought. In acute drought conditions hydrological and agricultural varieties are frequently endemic in certain areas in a chronic form needing long term measures rather than crisis management or emergency response required for an acute drought. Crisis management is most frequently required in case of meteorological drought although the crisis precipitated by an acute drought acquires extra-ordinary severity where hydrological drought is already in evidence.

An acute drought is an event due to scarcity of water, fodder or food, when rainfall is below 60 to 90 percent of normal rainfall or deficient rains of 20 to 59 percent below the normal rainfall, substantially event of failure of agricultural operations causing loss of rural livelihood and reduction in over all availability of water for human, livestock and crop consumption. Adverse effects of droughts over the population of the affected areas is multi-faceted. In economic terms it losses production of crops, dairy and livestock, timber and fisheries. In addition to human loss, drought has severe loss on national economic growth and development. Droughts have direct income loss to cultivation and indirect loss to agricultural labour, business and trade. The losses include deficiency in tourism inflow and other recreational business.

Power generation from hydro electricity is seriously affected due to lesser storage of rain water. Industrial production is also affected specially for units of processing of agriculture produce. Decline in food production is resulted in hike in prices of agriculture produce. The employment opportunities are mostly vanished and state government is compelled to defer or suspend collection of land revenue.

7.1 PRESENT SYSTEM OF DECLARATION OF DROUGHT AFFECTED AREAS

The behaviour of monsoon is usually erratic and uncertain in the state. Kharif production is dependent on the quantum and distribution of rainfall. The monsoon normally onsets in the first week of July and withdraws by the end of first fortnight of September. Thus the optimum moisture availability period varies from 50 days under normal conditions with 10-28 potential rainy days. The behaviour of monsoon is classified as under:

(a) Normal season with normal onset, cessation and distribution of monsoon.

(b) Delayed onset of monsoon.

(c) Normal onset but early withdrawal of monsoon.

(d) Normal onset and cessation but prolonged drought period in between inter-spell dry period.

(e) Flood and excess rains.

(f) Uneven distribution of rains.

These are possible situations of monsoon behaviour in the state and in some cases of delayed monsoon or early withdrawal condition, crop failure in eminent but at any such duration there may be heavy downpour causing famine and flood conditions simultaneously which is generally disbelieved by outsiders. The crop damage is the end result of all the uneven situations of monsoon behaviour except excess rains.

7.1.1 Measures for Assessment of Crop Damage and Scarcity Conditions

Girdawari is the regular phenomena of revenue administration system where timely assessments are made about the cropping pattern. This process is employed in Kharif and Rabi crops as assessment is carried out from sowing to harvesting. In case of Kharif crops, which is mostly dependent on monsoon, first girdawari is carried at the time of sowing of crops and monsoon behaviour is well known to the personnel engaged in this task. The girdawari process for assessment of area sown under different crops is done in total state where each Patwari records about sowing practices in terms of area and crop variety. This information is made available to

State Government through the Board of Revenue. The objective of first girdawari is to know the area sown as response to monsoon. The collected information is compared with normal trend and deviations are made in terms of percentage increase or decrease from normal conditions.

With first assessment the state government is able to know the likely production in the state under different crops. For this purpose the actual area sown under different crops is multiplied with average crop yield and first assessment is made, which remains very crude but estimation of the crop status in the state. Such forecasts are helpful in timely assessment of crop yield in the state. Girdawari of different crops is decided in case of each crops and varies between two to five in case of some sensitive crops like cotton. This procedure is applied in each state for crop and production assessment. In case of crop failure District Collectors are directed to make assessment of crop damage. Such directions are given in the beginning of September when crops are at the stage of maturity and assessment of damage is possible.

In response to the directions of the state government, District Collector of each district gears up district revenue team to make quick assessment of the crop status and at village level Patwari is expected to perform this role. Each Patwari has one to ten villages in its charge according to area and visits each village and records the crop damage in each field of the village. The area and crop is already available in the record of Patwari from crop sowing assessment and present task is to make eye-view of crop damage. Patwari records the crop condition of each field and this system is recorded from Khasra Number, which is code number of each operational holding known to patwari and cultivators.

Assessment of crop damage is done in percentage terms under three categories-crop damage below 50 percent, crop damage from 50 to 74 percent and crop damage from 75 to 100 percent. Each Patwari sends its first and final report to Girdawar and Each Girdawar sends consolidated reports to District Collector. The District Collectors on their part remain vigilant and send prior information of their respective districts to the State Government about the failure of monsoon. When the District Collector feels that scarcity conditions are developing and the distress is deepening, the report is submitted to state government as prior information of the district. The Girdawari report is submitted to the State Government with proposals for declaration of scarcity of any category giving full report covering the following points:

(a) Statement showing deficiency in rainfall as Proforma-I.

(b) Statement showing the water level in he various major irrigation and minor irrigation sources in each Tehsil in the district as Proforma-II.

(c) A statement showing the different categories of population affected by drought in Proforma-III.

(d) Position regarding drinking water supply as Proforma IV.

(e) Position of cattle and fodder with normal and prevailing prices as Proforma-V.

(f) A statement showing the number of aged persons, who are not in receipt of old age pension and who need economic support during the occurrence of drought in Proforma-VI.

(g) The comparative tehsil-wise statement of final Girdawari of crops in the area of Girdawari of crops during the preceding two years and damage of crops as Proforma VII & VII D.

(h) A statement showing the number of village artisans, category-wise, who are affected and who need economic support during the occurrence of drought in Proforma-VIII.

(i) A statement showing the number of children and nursing expectant mothers, who need special dieting supplement during the drought period in Proforma-IX.

(j) The nature of the principal crops grown normally, whether cash or food-crops and their relative extent.

(k) The estimate of crops in the area with the actual production during the two preceding years as percentage of loss or damage of crops.

(l) Area under different crops covered by the comprehensive crop insurance scheme.

(m) Whether the area is recognized as backward area.

(n) The opportunities of employment available on agricultural operations and normal Plan and Non-Plan works in the area.

The first report containing all above details is sent by 31st October and final report by 30th November. This is comprehensive report to be prepared by each District Collector about the drought condition in district and intensity of the drought is well addressed. Details of proformas as mentioned above are appended. The information about the drought affected areas to be submitted to the state government is quite exhaustive and all the district level departments are required to make quick assessment of the situation in respect of their respective fields. The effect drought in terms of its severity and spread is known from the field response in respect of various spheres, which are likely to be affected by drought conditions.

The information of severity of drought is respect of each affected district is prepared in statements given at (a) to (i) and further elaborated in view of the impact of drought over the population. Information in respect of items given at (j) to (n) is prepared in narration indicating the specific details. The purpose of this information is also to make assessment from respective sources of State Government resources and possible dovetailing from the Government of India schemes where additional funds can be managed as per their guidelines. At the state level, information received from all the scarcity districts is collected and assessment of scarcity in terms of area spread and requirement of funds is done to find ways and means for facing drought conditions in the ensuing months.

Similar exercise of rabi crop damage is also prepared on the basis of similar information from each district, where first report is submitted to the State

Government by 31 March and final report is submitted by 30th April. Damage of rabi crop is resultant impact of monsoon failure to severe extent and the ground water resources are not in a position to help in crop growth. Sometimes rabi crop is possible if winter rain occur in the state though chances of winter rains remain quite dim, but natural factors are governed by atmospheric conditions. Two spells of winter rains are sufficient for crops but its failure causes significant damage.

7.1.2 Submission of Scarcity Report to Government of India

Management of drought relief measures require huge funds which is not possible to manage by the State Government. Drought is a natural calamity and the Government of India helps state governments for which provisions has been made by the Finance Commissions. For drought relief, funds are primarily available from following sources:

(a) Calamity Relief Fund (CRF)

(b) National Calamity Contingency fund (NCCF)

(c) Additional allocation of funds from Ministries of Rural Development, Ministry of Agriculture etc.

(d) State own resources.

The Calamity Relief Fund has been constituted by the Government of India on the recommendations of the Ninth Finance Commission. It has been constituted for the purpose of financing natural calamity relief assistance. This scheme was first introduced in 1990-91 and continuing up to 2010. Funds remain in the accounts of the state government by transfer to fund in the ratio of 75:25 by the Central Government and State government respectively each year and remain deposited in the account. The purpose of maintenance of separate account and flow of funds on annual basis is to make the state governments strengthened to manage drought relief measures, if severity of the drought is possible to be met from the available funding.

On receipt of information from all the affected districts, the magnitude of severity and requirement of funds for initiating relief measure is well known to the state government. If the requirement of funds is possible to manage from the available Calamity Relief Fund, the state government takes decision in the cabinet to initiate relief activities. In case the requirement of funds is much beyond the available funds, the state government deems it proper to request the Government of India for funding. In such situation a memorandum is required to be prepared by the Disaster Management and Relief Department of the State Government indicating the severity of problem, affected districts, affected population and villages and damage caused. The requirement of funds need to cover all the aspects in proforma I to IX and other relevant details. The memorandum need to cover the state inability to manage drought relief measures and request for funding for the requisite amount beyond availability in CRF.

On receipt of the memorandum from the state government, the contents of the memorandum are examined in the Government of India and on satisfaction of the version of the state government in terms of the severity of the problem, constitutes a Central Committee to have first hand review of the severity of the problem and visits the state. The preliminary discussion is convened by the central team with the officers of the state government and team members visits various areas of the state to assess the severity of the scarcity conditions and also meet people of selected areas. Later the team returns to state headquarter to have concluding discussions. The central team prepares its recommendations and submits to the Government of India. Decision about the sanction of funds is taken on the recommendations of the central committee and state government is appraised with the decision and quantum of sanction of funds.

7.1.3 Declaration of Drought

The state government declares drought affected areas keeping in view the intensity of the problem and availability of funds from all the resources. The Ministry of Rural Development and Ministry of Agriculture also provide funds at the request of the state government, which is additive to the available resources. In addition the state government dovetails its departmental funds for the drought affected districts and areas. The reports of the district collectors is already available to the state government where intensity of various problems is already mentioned. At the state level two types of committees are constituted to take assessment of district problems. The apex level committee is constituted under the Chairmanship of the Chief Minister where ministers are assigned affected district to make assessment of problem and initiate relief measures in view of the severity of problem. The position is reviewed by the committee and instructions in respect of various programmes for crisis district is addressed suitably.

Officer level committee is constituted under the Chairmanship of the Chief Secretary or senior most officer of the state government where secretaries are assigned affected district to supervise the progress of relief measures and implementation of relief programmes. During district and field visits, the assigned secretary is required to assess availability of food grain, fodder, drinking water, medicines and nutrition support, assistance to old and infirm people. The respective district collector is instructed for acceleration of additional relief measures and submits report to the committee. Both these committees meet frequently and various decisions are taken in respect of increase of wage employment and labour ceiling to each district. These activities continue till occurrence of monsoon and all the affected districts need to ensure for no life loss for want of food and other essential commodities.

7.2 PROBLEMS OF SATELLITE INFORMATION SYSTEM

The satellite information system is most developed in the country and capacity of photo imageries has been enhanced to the level to get photo of land situations of one square kilometer of area. Even sub-surface information is possible to the be gathered. However, this information system is not being used in drought related issues for various reasons. Basic problem for non-acceptance of satellite information system in drought assessment is the fact that crop sowing is varied in any area with significant variation from early to late sowing and satellite photo imageries cannot distinguish between late crop and area sown. The crop sowing starts from occurrence of rains continued in marginal and degraded land if rainfall is sufficient. Thus the photo imageries are unable to distinguish between delayed crop and crop damage and no generalized information is possible to the taken on the basis of available information.

In addition, every state likes to exaggerate the severity of drought in terms of intensity and magnitude simply for the reason that the Government of India provides very scanty funds after getting the assessment by their own officers. Under the central state relations, the Central Government has wide powers and capacity to adjudge the scarcity conditions. The political shape is given to drought to extract more funds from the centre and benefit the people to excavate the electoral scenario. This position is not possible with the satellite information system. For more than three decades the drought has taken political shape to stress the centre to help respective state government posing the problem of drought. Though central funds are received to state government in the form of 30 percent grant and 70 percent loan to be repaid in phased manner.

With this approach the state governments are burdened with loan component but the political system looks the position from present and feels that burden of loans will be responsibility of the coming political party in the government. The burden of loan over the state government multiplies and the strategy remains to be sorted out by the next Finance Commission while making assessment of natural calamities in the state and the Calamity Relief Fund would be further enhanced. These are purely political and policy issues which cannot be met with satellite based information system and none of the state has preferred the system. The manual assessment has much flexibility, where some districts are given more stress in comparison to other districts and regional issues are possible to be addressed in declaration of the drought affected areas and relief measures.

7.3 RELIEF MEASURES AND THEIR EFFECTIVENESS

In critical scarcity conditions relief measures are carried out to provide wage employment to the rural people severely affected by drought. Wage employment is the direct employment method where construction activities are carried out in the affected villages and general activities carried out relate to construction of kachha roads, farm ponds, desilting of tanks or ponds etc. The people turning for employment are provided wages in the form of cash and food grain to meet their immediate requirements. Old and infirm people are given food grain without any employment treating as social cause. Fodder depots are established in affected areas to protect livestock. Drinking water arrangements are made through tankers in the villages where drinking water sources are extinct and people are facing hardship. Usually available water in medium and minor irrigation dams is used for meeting drinking water requirements in severe drought conditions.

Children below 6 years of age, lactating and pregnant women are provided nutrition support in the areas where no regular scheme of similar nature is under implementation. For control of diseases, camps are organized in village to provide immediate assistance to people suffering from seasonal diseases. These all efforts are made to save people from miserable condition. Similarly drought proofing measures are addressed through desert development programme, drought prone area programme, wasteland development, watershed management, reforestation in degraded forest areas, development of community land etc.

The drought relief measures and drought proofing measures are carried out in the state in most unsuccessful manner where no permanent asset creation could be possible in last 50 years. Except for construction of school buildings and hospital buildings, no visible works exist in total state which may help sustainable employment to the people. Desert development programme could not justify its implementation in the desert region as most of the sand dunes are still mobile and mountains and hills are denuded. Drought prone area programme was started for drought proofing measures but remained utter failure and drought intensity has been increased manifold in the affected districts.

Watershed development programme could not help in land treatment and increase of irrigated area even after huge expenditure. The activities selected under the programme are based on satellite imageries where run off water can be used for irrigation and moisture retention but no remarkable success could be achieved and most of the treated lands have become more degraded after completion of schemes. Distribution of food and fodder as drought relief measure is wastage of government food grains and creation of creative assets. Even the funds received from of India are in the form of 70 percent loan to be repaid along with interest, which imbalances the state economy. The total drought mitigation and proofing programmes have just helped the personnel engaged in the task and generally remained wasteful expenditure.

7.4 IMPACT OF DROUGHT PROOFING MEASURES

The implementation system of Disaster Management and Relief Department is most ambiguous in terms of use of scare money and relief to people. The relief measures had started from 1951 and continuing as drought is recurrent feature in the state. Every year, the emergency plans are prepared and amount is spent. There is no systematic strategy initiated as policy measure of the state government to address activities required in the state and no impact is visible in any form. The severity of drought is increasing regularly as vigilant from the severity of drought. Ground water level has reached to most critical level in total state where extinct of these resources is feared. This condition puts question mark over the huge expenditure incurred on watershed construction and other measure to charge ground water. Condition of forest areas is also visible as most deforested regions and no success could be achieved in the efforts of all the department engaged in this task.

There is no strategy for spending money received for drought relief measures and also drought proofing measures. The state government remains interested to create labour ceiling for drought activities and there are various orders issued from time to time but no efforts have been made to evaluate the fete of government funds spent on drought. The Disaster Management and Relief Department get all the schemes implemented through the agency headed by district collectors. Line department and Panchayati Raj Institutions have no care to spent the money without considering guidelines and policy of the state government. The relief to the people is limited to the extent that their should be no human loss from hunger. In this framework out relief arrangements are being performed.

7.5 CRITICAL ASSESSMENT OF DROUGHT MITIGATION STRATEGY

Last twenty years have been characterized by worsening of the drought impact on the population, which is evident that in coming years the vulnerability of drought will increase particularly due to development pressures, population increase, environmental degradation that could itself lead to climatic changes. The strong and recurrent link between climatic changes and drought has highlighted the importance of initiating a new way of analyzing the manner in which communities can cope with drought. This can be done by strengthening the coping mechanism of the community through empowerment of the community for drought preparedness planning; integrated drought risk management plans at the district level using the Geographical Information System (GIS) based risk and vulnerability mapping. Promoting short-term and medium term dynamic drought mitigation interventions that can ultimately contribute to the long term drought preparedness and mitigation plan at the grass root level are real necessity.

The Government has been taking up various initiatives to mitigate drought impacts through short term contingency plans and interventions, which can help the community to respond to their immediate needs. With increased frequency and recurrence of this phenomenon in the drought affected districts, it has become important that the focus for forthcoming years be shifted from crisis management to risk management. This is better done by analyzing different cause-effect of the problems, promoting user friendly early warning dissemination system. and enhancing the ability of the communities to cope with drought by creating awareness and enhancing knowledge on drought risk education options.

Wide range of physical and social factors such as demographic trends and geographic characteristics determine social vulnerability to drought, which require multi-sectoral approach. Therefore, one of the aims of the drought mitigation plan must be to combine the traditional and indigenous knowledge of drought preparedness and mitigation with applications of modern science in the community drought preparedness and implementation of contingency plans. In addition, efforts are required to guarantee that the affected communities are fully aware of the importance of detailed analysis of the past experience of drought years while deciding the cropping patterns for the monsoon and post monsoon season. Involvement and ownership of Panchayati Raj Institutions (PRIs) in the implementation of the project at the grass root level is to ensure sustainability of the planning initiatives intended for particular area or region. The approach lies within the vulnerability reduction and risk management framework, which is sustainable and continuous process as compared to the traditional preparedness focused on the drought response system. Keeping in view the above strategy, the long term plan must focus on:

(a) Awareness generation among the communities, specially among women and other vulnerable groups, in order to enhance their ability to predict, face and cope with drought like situation.

(b) Strengthening community based initiatives and promoting community based drought preparedness and mitigation measures through PRIs in the most critical districts of drought prone regions, ensuring that women are a crucial and participant in the development and implementation phases of the drought mitigation plans.

(c) Training and capacity building of the state institutions and departments through increased involvement of academic institutions.

(d) Strengthening the existing administrative, legal and institutional frameworks at state level, district level and grass root levels for drought risk management.

7.5.1 Drought Mitigation Strategy

The drought is most severe disaster affecting rural population of affected regions to significant extent. The drought condition can be converted in to famine due to delayed relief measures and assessment of grass root problems in necessary before addressal. The strategy of drought mitigation measures needs to cover the issues, which affect most of the people of the area. People must realize that their destructive acts of environmental degradation need to be stopped with collective efforts as these issues play vital role in erratic and uncertainty of monsoon and drought is the resultant impact of human acts. Development and destructive activities cannot work together and all the environmental degradation measures need to be controlled for their long term benefits. Drought risk management strategy need to tackle drought as a complex phenomenon through an end-to-end approach and take in to consideration the issues the people suffer most. Drought management plan cannot be universal for total state but area and region specific issues need to address following issues:

(a) Existing coping mechanism and indigenous knowledge on drought proofing.

(b) Development of community based drought preparedness and mitigation plans ensuring women's concerns and drought related issues to be taken into consideration in the regular development and in its implementation phase.

(c) Linkages between the community based drought preparedness and mitigation plans and regular activities for desert development programme and drought prone area programme need to be carried out in effective manner.

(d) Design and development of drought information and knowledge management systems using integrated combating technology at state district, block and grass root levels.

(e) Review and strengthening of existing administrative and managerial arrangements at state and district levels.

(f) Sustainable drought risk reduction strategies at state, district, block and grass root levels.

(g) Coordination between different departmental activities being carried out in the state need to be addressed in result oriented manner.

(h) There is necessity to initiate drought relief measures as part of long term sustainable development strategy.

7.5.2 Institutional Framework and Capacity Building at State Level

There is necessity to establish a state drought information management cell in the Disaster Management & Relief Department for management of drought. This cell must gather historical database of all climatological factors and need to be equipped with appropriate information technology base system to undertake trend analysis to feed into the drought preparedness plans for each drought year. State level resource inventory, online farmers' advisory services based on weather forecasts , inventory of district risk drought management plan etc should be developed and managed by these centers. Linkages with remote sensing institutions for monitoring drought proofing initiatives need to be established. The state level Disaster Management and Relief Department must develop its capabilities to make quick assessment of drought situation by state task force, preparation of drought risk management plans, training of damage and needs assessment and comprehensive contingency plans to implement quickly and provide relief to the people. State level consultations need to be organized to seek opinion of important stakeholders for drought reduction strategy.

7.5.3 District Drought Risk management Planning

At the district level, focus must be on awareness generation among community leaders and members of Panchayati Raj Institutions, training and capacity building of government officers and officials associated with drought management plans and decision support mechanism. Advance information system need to be developed in each district on GIS platform is also helpful for conducting drought risk modeling and planning for different drought interventions, which should be implemented as part of district drought preparedness plan. The planned interventions must focus on alternate cropping pattern of agricultural crops, selection of drought adapted fodder varieties, water resource allocation. These units need to be associated with the information dissemination of early warning signals like in coastal regions. Additionally, this programme must focus to provide early warning signals on the basis of rainfall data and drought risk models to enable the district administration and local self-government institutions to initiate advance measures like alternative cropping pattern, provisioning of drinking water through better conservation and management of run-off, stockpiling and promotion of food and fodder cropping for livestock in most vulnerable areas and stockpiling of food grains as well as identifying need for wage employment schemes in the areas which is likely to experience more insecurity due to increased livelihood options. Often, these may be areas prone to distress migration and with early signals on drought situation, advance planning could be initiated on the basis of district drought risk management plans. All the extension services of agriculture and animal husbandry departments need to be oriented towards drought preparedness planning. These interventions must aim at improving the efficiency of government drought risk management activities at district and sub-district levels through their capacity building.

7.5.4 Community Based Drought Preparedness and Mitigation Planning

Ideal location for such activities is the village where activities must be aimed to community awareness generation, development of community based drought preparedness and mitigation plans, training and capacity building of village volunteer task force to perform various tasks related to water conservation, alternative crop planning, management of cattle, care of vulnerable groups of society etc as well as promotion of alternative livelihood support activities with active participation of women. These community based drought preparedness and mitigation plans need to be prepared by the community and need to be approved in the Gram Sabha and to become an integral part of annual village development plan, feed into block plans and form integral part of district development plan.

During the planning process, linkages between community based preparedness and mitigation plans and other mitigation initiatives like desert development programme and drought prone area programme and other area based programmes need to be dovetailed with these plans to match their requirements. All the development activities need to be managed in he manner to address village need based activities at village level. Generally there are multiple schemes where funding is available through different sources and departmental personnel manage their implementation without consulting the village people where these schemes are to be implemented. If all the departmental activities meant for rural areas are able to address the village people requirement, half of the problem can be solved automatically. There is flexibility in departmental schemes to fulfill village requirements and if there is any problem, that can be addressed through state level management.

7.5.5 Implementation of Community Based Drought Mitigation Plans

Any plan intended to be implemented by the community need to have flexibility and initiated as demonstrative plan where drought occurrence had worst impact for consecutive years. In Rajasthan, there is acute shortage of water and such programme need to address rain water harvesting as per the experience of the community as per feasibility of soil structure, where such measures may help in increase ground water level or may be helpful for irrigation purposes. Such plans must cover water harvesting even in village by roof water harvesting, cropping pattern with water saving devices, development of common property resources. A detailed village resource mapping with active participation of community can be helpful in strategy making for community based drought mitigation plan.

Pilot project must be implemented through women self help groups, water user groups or village forest committee. If community based organization is actively shoulder the responsibility of preparation and implementation of plan, such organizations need to be encouraged to tackle village problems of drought related issue or combating drought mitigation measures. The activities may include initiatives of early warning system, community awareness programme, rain water harvesting, planned drainage to reduce water logging, preparation of dynamic agriculture or livestock contingency plan, soil and water conservation, community nurseries for Afforestation, art and craft development and other livelihood options based on village demand based activities.

Oran culture is ancient method of community land development in desert and tribal areas of Rajasthan, where people of village protect their vegetated land and do not allow village people or outsiders to harm trees. Jaisalmer district has 27 orans spread over in 43300 hectare area and single oran of Bhadaria village is situated in 15000 hectares. Similarly Barmer district has 253 orans in about 50000 hectare area and Kundla village oran has area of 7500 hectares. This is the ancient culture which need to promoted in other areas through community participation, which can help in maintaining the environment and ecological balance of the areas.

Proforma-I

RAINFALL DATA (Rainfall in mm)

S. No	**Name of Tehsils**	**JUNE**				**JULY**			
		No. of rainy days	Actual rainfall	Normal Rainfall	Variation (+ or -)	No. of rainy days	Actual rainfall	Normal Rainfall	Variation (+ or -)
1	**2**	**3**	**4**	**5**	**6**	**7**	**8**	**9**	**10**

Table Continued ...

S. No	Name of Tehsils	AUGUST				SEPTEMBER			
		No. of rainy days	Actual rainfall	Normal Rainfall	Variation (+ or -)	No. of rainy days	Actual rainfall	Normal Rainfall	Variation (+ or -)
1	2	11	12	13	14	15	16	17	18

Table Contiued ...

S. No	Name of Tehsils	TOTAL FOR THE SEASON				PERIOD OF DRY SPELLS IN THE SEASON
		No. of rainy days	Actual rainfall	Normal Rainfall	Variation (+ or -)	
1	2	19	20	21	22	23

Proforma-II

IRRIGATION SOURCES AND WATER LEVELS

S. No	Name of District	Name of Reservoir/ Tank/ other source (specify)	Total Capacity		Present Water level in the week		Normal water level during the same period of a normal year		Period up to which the water in the source will be sufficient
			Gauge/ RL Feet	Capacity MCFT	Gauge/ RL Feet	Capacity MCFT	Gauge/ RL Feet	Capacity MCFT	
1	2	3	4	5	6	7	8	9	10

Note: The information in the above table should be given for all the irrigation sources for which gauging are available. In respect of the other Minor Irrigation sources, where no gauging arrangements are available, the adequacy of the water in the sources shall be furnished in a separate note along with the table.

Proforma-III

POPULATION AFFECTED BY DROUGHT

S. No	Name of Tehsil	No. of villages	No. of total house holds	Total Popu-lation	Small Farmers		Marginal Farmers		Agricultural Labours	
					No. of house hold	No. of persons	No. of house hold	No. of persons	No. of house hold	No. of persons
1	2	3	4	5	6	7	8	9	10	11

Table continued ...

S. No	Name of Tehsil	No. of farmers having single crop area		Number of persons Affected due to drought				Is there migration of people in search of work
		No. of house hold	No. of persons	Small farmers	Marginal farmers	Agri. labours	Farmers having single crop area	
1	2	12	13	14	15	16	17	18
	Total							

Note: 1. The number of persons affected due to drought from among SF, MF and Agricultural labour population have to be calculated with reference to the drought index arrived at the tehsil. If 25 is the drought index for tehsil, 25% is total under SF, MF and AL will be taken as affected due to drought.

2. The drought index is defined as average of the % of the deficiency in rainfall and % of area affected due to adverse seasonal conditions from the normally sown area. The area affected is the total of area not sown/ transplanted from the season and area sown/ transplanted but withered of affected for want of rain.

3. Even the rainfall is in excess of normal, the crop may be damaged due to dry spells in which case the drought index should be computed as if the deficiency in rainfall is zero. In such case the drought index will be half of the % of area affected due to adverse seasonal conditions.

4. Farmers having single crop area are those who are having holding more than marginal farmers.

Illustration: The percentage deficiency in rainfall is the % of col 18 to col 17 in proforma-I when the variation is (-). Let this be 'A'.

The percentage of area affected due to adverse seasonal conditions is the % of Col. 8 to Col.5 against the total for the tehsil in Proforma-VII, let this be 'B'. The drought index for the tehsil is (A+B).

Proforma- IV

DRINKING WATER POSITION

S. No	Name of Tehsil	Total No. of villages	Present Position regarding drinking water facilities					
			Bore Well		Hand Pumps		Open Well	Open well
			Total no. existing	No. in use	Total No. existing	No. in use	Total existing	No. in use
1	2	3	4	5	6	7	8	9

Table continued ...

S. No	Name of Tehsil	Present position regarding drinking water facilities							
		PHED protected water supply schemes		Surface source of drinking water		No. of villages not covered by PHED schemes		No. of villages/ town where drinking water scarcity is felt	
		Total existing	No. in use	Name	No. of villages/ towns where existed	Villages	Habita-tions	Village	Town
1	2	10	11	12	13	14	15	16	17

Table continued ...

S. No	Name of Tehsil	Augmentation Proposals							
		No. of open wells that could be descended	No. of bore wells that could be repaired and put into use	No. of new open wells to be dug	No. of new bore wells to be dug	No. of hand pumps to be dug	No. of villages for which water should be supplied by tankers		No. of hand pumps to be repaired
							Villages	Towns	
1	2	18	19	20	21	22	23	24	25

Proforma-V

CATTLE AND FODDER

S. No	Name of Tehsil	No. of famine affected villages	No. of cattle	Quantity of dry fodder required for them in the season	Quantity of fodder available locally	Deficit in fodder stocks existing	Are there any migrations of cattle for fodder and pasture	Are there any losses of cattle due to drought conditions	Prevailing price of fodder
1	2	3	4	5	6	7	8	9	10

Proforma-VI

DESTITUTES - GRATUITOUS RELIEF

S. No.	Name of Tehsil	No. of villages	No. of old and infirm people	No. of persons receiving old age pension	No. of persons who require gratuitous relief	Are there any starvation deaths reported? If so the number
1	2	3	4	5	6	7

Proforma -VII

CULTIVATION PARTICULARS

S. No	Name of Tehsil	Name of principal crop	Actual area cultivated in the season	Normal area under this crop in a normal year	Area left uncultivated	Extent of crop damaged after sowing/ transplanted due to failure of rain
1	2	3	4	5	6	7

Table continued ...

S. No	Name of Tehsil	Total Loss (%)			Total Loss	Loss of production in Tonnes	Approximate value of the loss of production in Lakh Rs.
		Below 50%	50% to 75%	75% to 100%	Col 9+10		
1	2	8	9	10	11	12	13

Note: 1. Normal Area: Col 5 The average of preceding 5 years

2. Loss of Production: Col. 12: This is product of Col. 11 and average yield – average of preceding 5 years yields.

Proforma -VII D

DAMAGE REPORT OF KHARIF/ RABI CROP

YEAR .... SAMVAT ....

S. No	Name of tehsil	Total villages	No. of affected villages	Population of affected villages	Total geographical area in Ha.	Area of affected villages		
						Cultivated area in Ha.	Sown area in Ha.	Area of damaged crop in Ha.
1	2	3	4	5	6	7	8	9

Table Continued ...

Percentage of crop damage			Produ-ction in MT	Value of damaged crop (Lakh Rs.)	Land revenue of affected villages (Rs)	Likely suspension of land revenue (Rs.)	Total cattle popula-tion (lakh)	No. of affected cattle (lakh)
Less than 50%	50-74%	75-100%						
10	11	12	14	15	16	17	18	19

Proforma- VIII

VILLAGE ARTISANS

S. No	Name of Tehsil	No. of villages	Village artisans (Category-wise)	No. of households	No. of persons in these households	No. of persons affected by drought
1	2	3	4	5	6	7
			1			
			2			
			3			

Proforma- IX

SPECIAL NUTRITION FOR CHILDREN, NURSING AND EXPECTED MOTHERS

S. No	Name of Tehsil	No. of villages	No. of children below 6 years of age	No. of children of poorer sections who should receive dietary supplement during drought	Total number of pregnant women	No. of pregnant women of poorer section who should receive help	Total number of nursing mothers	No. of mothers of poorer sections who should be assisted
1	2	3	4	5	6	7	8	9

Table continued ...

S. No	Name of Tehsil	No. of children already covered under ICDS		No. of pregnant mothers covered under ICDS		Balance to be covered		
		No. of villages	No. of children	No. of villages	No. of pregnant women	No. of villages	No. of children	No. of pregnant women
1	2	10	11	12	13	14	15	16

8

STUDIES FROM DIFFERENT AGRO-CLIMATIC ZONES OF DROUGHT AFFECTED AREAS

Rajasthan state is seriously prone to drought conditions, irrespective of its geo-physical and climatic conditions. The vulnerability of drought is common to all the districts of hyper-arid, arid and semi-arid regions of the state with varied frequency. Drought is a natural factor and its occurrence is dependent mainly on failure of monsoon, which is erratic and uncertain. The state has different rainfall conditions varying from 18.55 cms in Jaisalmer to 95.03 cms but the frequency of drought has probability of occurrence in every district as evident from past records. The rainfall adaptability in different regions is well versed to the geo-physical conditions of different regions. There are various districts prone to drought conditions while there are also few districts which are gripped with drought conditions once in eight years. This is the impact of location and physiography of those regions.

8.1 AGRO-CLIMATIC ZONES

The climatic conditions of various regions affect the agricultural cropping patterns and different areas produce different crops. Amongst a host of climatic factors, rainfall, temperature, humidity, wind velocity and duration of sunshine etc affect the cropping pattern in most significant manner. Annual rainfall and its distribution over the entire year, regimes of diurnal and annual temperatures are the prominent factors affecting agriculture and the life style of the people. On the basis of climatic conditions and agricultural produce, Rajasthan state has been divided into nine agro-climatic zones and each one having special characteristics of its own varied with others. Agro-climatic zones have been delineated on the basis of geo-physical conditions and natural factors as actors for ecological and environmental conditions. The impact of drought remains varied over each of the agro-climatic zone.

8.1.1 Arid Western Plain (Zone I-A)

This region comprises of Bikaner, Jaisalmer and Barmer districts, Phalodi, Shergarh, Osian and Jodhpur tehsils of Jodhpur district and Dungargarh, Ratangarh and Sardarshahar tehsils of Churu district. This zone is most arid part of the state where the annual rainfall varies from 10 to 40 cms, quite often erratic to the extent that the entire annual rainfall may fall on a single day and the rest of the year may remain dry. Summer temperatures are always at extremes where diurnal range may exceed even 20 degree C. During the day, the summer temperatures may be as high as 49 degree C, but in the night, the temperatures usually fall to less than 20 degree C. In winters, the day temperatures are highest but the night temperatures may be near freezing point. Winters are of short duration not exceeding two months-December and January.

This is an area of not-so-developed drainage system where there are no flowing streams. Owing to poor rainfall, surface water resources do not exist while ground water resources are often deep and brackish. Natural vegetation is therefore, only seasonal. With the first showers in late july, a few grass species grow and shrubs and dwarf trees become green. But soon after retreat of the monsoon in mid-September, the vegetation dries up, leaving only a few perennial shrubs and a thin pad of pale grass on the landscape. Mostly rain-fed crops like bajra, kharif pulses, guar etc are grown during the kharif season. Rabi crops like wheat, rape seed and mustard are grown only in areas where ground water is available for irrigation.

8.1.2 Irrigated North Western Plain (Zone I-B)

Entire Ganganagar and Hanumangarh districts which are alluvial and aeolian plain, formed by the Ghaggar, the ancient river Saraswati, forms this agro-climatic zone. The Ghaggar channel, even though 7 kms wide at some places, unindates nearby villages, during heavy rainfall in its catchment areas. A part of this region, which is arid, is the northern extension of the Indian Thar desert covered with wind blown sand. Extreme aridity, marked with high summer and low winter temperatures, is its unusual climatic characteristics. The average annual rainfall is about 40 cms. High speed winds blow in summer.

The area is rich in agriculture produce on account of a well developed system of canal irrigation. A large network of Gang Canal and Bhakhra Canal which along with the Indira Gandhi Canal has made the entire area green and productive. Due to abundance of canal water irrigation, this region has become the greenery of Rajasthan. Among the kharif crops cotton, sugarcane and pulses are important. In rabi season, wheat, mustard, gram, vegetables and fruits are produced. The total production as well as the productivity levels of all crops are relatively much higher as compared with other parts of the state.

8.1.3 Transitional Plains of Inland Drainage (Zone II-A)

This zone comprises of entire Nagaur, Sikar and Jhunjhunu district, Taranagar, Churu and Rajgarh tehsils of Churu district. The area is covered with sand dunes and inter-dunal sandy plains. Drainage is not well developed and streams which flow in the rainy season, disappear in sandy fields after covering some distance. Climatically, this zone is slightly better as compared with the adjoining zone of Arid Westerly Plains. Rainfall is slightly higher, temperatures in summer months become very high, while winters are very cold. Irrigation is restricted to areas with good ground water potential. Bajra, sesamum and kharif pulses are the main crops of the rainy season. Wheat, barley, mustard and gram are grown as irrigated crops or on conserved soil moisture during rabi season. This part has very limited ground water resources at significant depth with brackish water.

8.1.4 Transitional Plain of Luni Basin (Zone-II-B)

This area lies between the Aravalli ranges and western arid region. The region encompasses entire districts of Jalore, Pali Reodhar and Sheoganj tehsils of Sirohi district and Bhupalgarh tehsil of Jodhpur district.The region has semi-arid climate with an annual rainfall of 30 to 50 cms. it is drained by the river Luni, which is seasonal river and flows only during rainy season. A number of paleo-channels also exist in this area. The western part of the region is dotted with sand dunes, interspersed in alluvial soil. Luni and its several tributaries like Sukri, Mithri and Jawai have made this area productive. The climatic conditions are almost the same as in the western arid region except that the rainfall is slightly higher. The ground water level is high in the river basins and has been usefully tapped for irrigation. Vegetation is xerophytic and sparse in he western part but in the east and on the slopes of the Aravalli ranges, there is mesophytic vegetation in the form of woodlands, open forests and grasslands. The area produces bajra, maize, guar, sesamum and pulses in the kharif season, while wheat, barley and mustard are the dominant crops of rabi season, specially in irrigated areas.

8.1.5 Semi-arid Eastern Plains (Zone III-A)

This region comprises of four districts viz Jaipur, Dausa, tonk and Ajmer. River Banas with its several tributaries forms a rich fertile plain. On the western side, the region is flanked by the low Aravalli hills, which extend from the south-west to the north-east. The annual rainfall of the region varies from 50 to 60 cms with an increasing trend towards the east. Summer and winter temperatures are not as extreme as in the arid west but the summer temperature may reach around 45 degree C and in the winter, minimum temperature remains up to 8 degree C. The water table varies from 15 to 25 metres but annual fluctuations are high, specially in the years when the south-west monsoon fails and the yearly replenishments are low. Surface water sources are scarce and so harnessing of the ground water resources has been going on an accelerated pace. Natural vegetation is of the mixed zerophytic and mesophytic type, but owing to heavy felling of trees, the surface mantle has been robbed of its natural wealth. In the total gross cultivated area of this zone, bajra, sorghum and pulses are grown in the kharif season and wheat, barley, gram, mustard are main rabi crops. Productivity of all crops in this zone is better than that of the agro-climatic zones that are to the west of the Aravalli range. This zone is greatly disturbed with the alkalinity and other land degradation features due to excess draft of ground water.

8.1.6 Flood Prone Eastern Plains (Zone III-B)

This reason comprises of the districts of Alwar, Bharatpur, Karauli and Dholpur, Mahua tehsil of Dausa district, Gangapur, Bonli and Bamanwas tehsils of Sawai Madhopur district. Except of few low hills, which exist in Alwar, Karauli and Sawai Madhopur districts, the entire region is flood plain of Banganga and Bhambiri rivers. In periods of heavy rainfall, the rivers outflow their banks and inundate the surrounding villages. Climatically, the area is similar to the plains of Banas seasonal river, but the rainfall is relatively higher in the east and annual average being about 75 cms. Natural vegetation exists on mountain slopes, wetland areas and protected zones but the excessive plundering of forest wealth has dwindled the natural cover. The region produces a variety of crops because both the surface water and ground water irrigation sources exist. A network of canals drawn from the upper Yamuna Canal and the Panchana Dam irrigate this area. The ground water aquifers vary from 5 to 15 metres and accordingly well water irrigation is also practiced. The region produces bajra, sorghum, maize, sugarcane, sesamum and a variety of pulses in the kharif season. Wheat, barley, gram and mustard are dominant crops of kharif season.

8.1.7 Sub-humid Southern Plains & The Aravalli Hills (Zone IV-A)

Dhilwara district, all tehsils of Udaipur district except Dhariyawad, Salumbar and Sarada, all tehsils of Chittorgarh district except Chhoti Sadri, Pratapgarh, Arnod and Bari Sadri, Abu Road and Pindwara tehsils of Sirohi district form this agro-climatic zone. For the most part, this is a region of low Aravalli hills with inter-mountain plateau, deeply dissected by streams and rivers of seasonal character. The region has moderately warm climate in summers and with warm winters. The annual rainfall varies from 50 to 95 cms. in Mount Abu region. The highest precipitation in the state is recorded in Abu hills of Sirohi district. There are number of surface water streams like Ghambiri, Sabarmati, Banas and its tributaries but they are all ephemeral.

The surface rocks are granitic and highly metamorphosed and so the depressions on the surface area filled with rain water in the form of natural tanks. The area is rich in natural vegetation which grows on the slopes of the Aravallis and in the wetland areas but excessive felling of trees has degraded these open forests. Tank water irrigation is most common where surface strata is hard structured. The area produces maize as chief food crop of the kharif season but in irrigated areas, paddy is also grown. In the rabi season, wheat, gram, oil seeds are main crops. In the areas of black soil, cotton and opium are also produced. Tube-wells have depleted the ground water to serious extent.

8.1.8 Humid Southern Plains (Zone IV-B)

Complete districts of Banswara and Dungarpur, Dhariyawad, Salumbar and Sarada tehsils of Udaipur district, Chhoti Sadri, Bari Sadri, Pratapgarh and Arnod tehsils of Chittorgarh district are included in this region. The area comprises of low

Aravalli hills with instrusions of black lava rocks. This is mostly tribal region where Bhils, Girasiyas and Damors live amidst forests and hills. The area comprises of low Aravalli hills with intrusions of black lava rocks. The area has humid climate with an average rainfall more than 70 cms per year. The temperature regimes do not fluctuate much in summer and winter so the area has mild winters and mild summers. The humidity is always high and all these factors, combined together, promote a profuse growth of natural vegetation. The mountain slopes have an abundance of natural forests.

There are a number of surface water streams mostly active during rainy season. Mahi and its tributaries like Anas, Arau and Jhakham have made this area very fertile. The recent commissioning of Mahi Bajaj Sagar multipurpose project has provided this area with canal water irrigation and electric resources. Cotton and sugarcane are the chief cash crops grown in the black soil region. Maize, sorghum and paddy are the chief food crops of the kharif season while groundnut, mustard, sesamum and rape seed are also grown. The thick forest area in hilly and rocky regions has suffered to great extent due to uncontrolled felling of trees, which has deteriorated natural environmental conditions and soil erosion has become regular feature.

8.1.9: Humid South-Eastern Plains

Popularly known as the Hadauti plateau, this region includes the districts of Kota, Baran, Bundi and Jhalawar, Khandar and Sawai Madhopur tehsils of Sawai Madhopur district. The black soil region of this plateau is fertile and is used for cultivation of sugarcane, cotton and opium. There are low hills of Gwalior series, interspersed with broad plateau of Vindhyan rocks. Large number of rivers drain this area. The Chambal is the main river along with its main tributaries like Parvati, Kali Sindh, Parwan and Banas. The development of canal irrigation system with a series of dams and barrages on the Chambal has made this area rich in agricultural production. Gandhi Sagar, Rana Pratap Sagar and Jawahar Sagar Dams together with Kota Barrange are hydro electricity generation resources and canal water irrigation. These systems were constructed to provide electricity and water for meeting the requirements of this region but in view of series of medium and minor irrigation dams in the neighbouring Madhya Pradesh state, the units are unable to get enough water for power generation and irrigation needs. Thus people of area are deprived of the facilities from created assets.

The region has warm summers but mild winters and the summer temperatures sometimes touch 45 degree C. The relative humidity remains generally high in this zone which is adaptable for vegetation germination and growth. The annual rainfall varies from 60 to 85 cms. Natural vegetation exists in the form of woodlands, parklands and open forests which have mostly degraded or extinct due to uncontrolled cutting of trees and bushes. Paddy and sorghum are the chief food crops grown in the kharif season. This area is suitable for soyabean crop also. Wheat, barley, gram and mustard are grown in winter season.

8.2 DROUGHT 2002 IMPACT IN AGRO-CLIMATIC ZONES

The drought of 2002 caused by the failure of the south-west monsoon in all the 32 district of the state was the worst in last 136 years in terms of severity and magnitude. The over-all rainfall deficit was- 64% which was also lowest in last 136 years. All the 32 districts received scanty and deficit rainfall. About 4.48 crore human population and 4.52 crore livestock population in 40990 villages were in the grip of drought of rare severity. As against the normal sown area of 129 lakh hectare in kharif, actual area sown was 82.26 lakh hectare but due to long dry spell throughout the month of July, 2002 almost entire crop was damaged incurring a loss of Rs. 4031 crores. During the rabi season, 40.73 lakh hectare area was sown against normal sown area of 82.13 lakh hectares. Ground water depletion was most severe ranging from 3.2 to 6 metres. Reservoirs received 32% water of their respective designed capacity and most of the surface water sources were dried up. There was unprecedented shortage of fodder and drinking water as 2002 drought was the fifth year of drought in succession.

8.2.1 Drought of Rare Severity

Drought 2002 was the fifth year of recurrent drought and its rare severity was the resultant impact as there was no time to recoup from the drought situation. Out of 237 blocks in the state, 74 blocks were continuously gripped by drought and the remaining blocks faced drought for the fourth year in succession barring year 2001-02. Severity of drought has been depicted in Table 8.1.

Table 8.1: Comparative Intensity of Drought 1998-99 to 2002-03

Particulars	1998-99	1999-00	2000-01	2001-02	2002-03
No. of districts affected	20	26	31	18	32
No. of villages affected	20069	23406	30583	7964	40990
Affected population (in lakhs)	215.07	261.79	330.41	69.70	447.80
Livestock population affected (in lakhs)	295.78	345.60	399.69	69.73	451.63
Crop damage (a)Area in lakh ha (b)Value in crore Rs.	 64.96 2283.49	 78.18 3407.02	 89.47 3511.77	 26.53 1252.27	 117.00 4414.00
Rainfall deficit	-3%	-16%	-29%	-5%	-64%
No. of irrigation tanks filled to full capacity	277	308	131	500	16

Source: Disaster management and Relief Department

The severity of drought was caused due to failure of monsoon in the state in totality and also when the crop germinated required water for growth. This position has been analyzed on fortnightly basis for the state as a whole from rainfall data on district average basis, the progress of monsoon rainfall in the state along with its deviation from the occurrence in normal conditions. The actual rainfall has been calculated on the basis of actual district arithmetic mean rainfall of 32 districts. The position of fortnightly rainfall in comparison to normal rainfall in respective duration is given in Table 8.2.

Table 8.2: Fortnightly Monsoon Analysis (June- September, 2002)

Fortnights Rainfall duration	Average Rainfall		
	Normal (mm)	Actual (mm)	Deviation (%)
1.6.2002 to 15.6.2002	23.9	6.7	(-) 71.8
16-6-2002 to 30-6-2002	23.9	60.4	(+) 153.0
1-7-2002 to 15-7-2002	91.4	2.4	(-) 97.0
16-7-2002 to 31-7-2002	97.5	10.9	(-) 88.8
1-8-2002 to 15-8-2002	93.9	68.3	(-) 27.3
16-8-2002 to 31-8-2002	100.2	39.0	(-) 61.1
1-9-2002 to 15-9-2002	50.2	38.1	(-) 24.0
16-9-2002 to 30-9-2002	50.2	4.0	(-) 92.0
Total position 1-6-2002 to 30-9-02	**229.8**	**531.2**	**(-) 56.74**

Source: IMD, Pune

The analysis of occurrence of rainfall during monsoon months reveals that during the period 16-6-2002 there was widespread rainfall more than 153 percent of normal rainfall in the duration and significant number of cultivators were tempted to sow crops in their cultivated fields. July month was severely dry which damaged almost all the crops due to failure of monsoon in this duration. The germination process started in June could not be continued and damage was severe. Cultivators awaiting the rainfall in July were disheartened due to failure of monsoon and deferred sowing process. Thus seeds grown in early sowing period damaged as July month remain most dry to normal rainfall in the duration.

The drought of 2002 became disastrous famine due to consecutive failure of monsoon in successive five years and in this duration the economy of the people and the state was completely dwindled. The district wise position of drought was more drastic as some areas were worst affected by the hazardous impact. The unprecedented situation of drought became major challenge for the state government and the people of the state as there was necessity to initiate relief measures to drought affected areas. Funds available under the Calamity Relief fund were too short to meet the requirements and the financial position of the state government was precarious to manage the calamity of gigantic magnitude. Therefore, the state government prepared memorandum for assistance from the Central government, where the district was severity of drought conditions was narrated, which is given in Table 8.3 and Figures 24, 25 and 26.

Table 8.3: Extent of Scarcity of Drought 2002 (*Samvat* 2059)

District	Total Villages	Affected Villages			Affect. popu-lation	Sown area	Affected area	Affected Live-stock	Crop damage	Susp. of land rev.	Rain-fall deficit
		50-74 %	75-100 %	Total	In lakh	Lakh ha.	Lakh ha.	lakhs	Crore Rs	Lakh Rs	in % (-)
1	2	3	4	5	6	7	8	9	10	11	12
1. Ajmer	1076	11	1065	1076	11.78	3.06	2.97	7.67	184	16	65
2. Alwar	1984	123	1857	1980	27.65	2.45	2.16	15.41	129	33	72
3. Banswara	1524	1510	—	1510	13.49	2.23	1.30	11.82	77	9	38
4. Baran	1233	688	519	1207	13.76	1.19	0.91	6.73	87	17	62
5. Barmer	1975	1	1974	1974	14.35	9.77	9.67	42.02	220	—	69
6. Bharatpur	1505	—	1504	1504	19.25	1.83	1.70	7.96	85	22	50
7. Bhilwara	1788	290	1494	1784	14.71	3.20	3.03	19.77	203	38	56
8. Bikaner	889	—	889	889	19.01	1.24	1.01	23.17	10	13	85
9. Bundi	870	369	501	870	8.86	1.60	1.23	9.62	87	10	56
10. Chittorgarh	2415	1193	1151	2344	16.34	3.89	2.80	17.16	203	22	51
11. Churu	907	12	895	907	13.73	4.32	4.24	18.10	121	—	64
12. Dausa	1061	117	939	1056	11.93	1.44	1.28	6.56	59	9	62
13. Dholpur	828	3	825	828	8.87	0.71	0.59	4.53	43	5	44
14. Dungarpur	872	—	872	872	8.74	1.09	0.93	9.58	9	1	37
15. Ganganagar	3028	484	1668	2152	7.05	1.53	1.13	8.80	2	27	76
16. Hanumangarh	1912	314	1525	1839	11.50	3.29	2.79	8.98	—	—	79
17. Jaipur	2340	35	2237	2272	26.16	4.89	4.37	20.36	275	51	71
18. Jaisalmer	688	36	607	643	4.43	0.24	0.22	12.56	4	3	83

District	Total Villages	Affected Villages			Affect. popu-lation	Sown area	Affected area	Affected Live-stock	Crop damage	Susp. of land rev.	Rain-fall deficit
		50-74 %	75-100 %	Total	In lakh	Lakh ha.	Lakh ha.	lakhs	Crore Rs	Lakh Rs	in % (-)
1	**2**	**3**	**4**	**5**	**6**	**7**	**8**	**9**	**10**	**11**	**12**
19. Jalore	728	5	723	728	11.43	4.44	4.30	6.26	260	44	59
20. Jhalawar	1614	325	1266	1591	10.70	2.68	1.47	8.62	114	8	46
21. Jhunjhunu	867	—	867	867	16.00	2.32	2.27	12.43	76	4	75
22. Jodhpur	1082	36	1046	1082	24.06	1.02	—	33.07	280	—	76
23. Karauli	829	9	820	829	9.73	1.31	1.23	6.38	82	17	69
24. Kota	947	521	379	900	9.05	1.28	0.88	3.73	139	35	53
25. Nagaur	1508	34	1474	1508	19.49	5.87	5.40	19.57	381	—	69
26. Pali	973	8	965	973	14.76	3.23	3.16	28.04	301	—	68
27. Rajsamand	1004	10	994	1004	8.03	0.86	0.76	11.93	81	10	49
28. S. Madhopur	800	62	738	800	11.16	1.05	0.93	8.18	51	—	71
29. Sikar	1017	6	1011	1017	18.45	3.80	3.60	16.63	198	10	67
30. Sirohi	474	—	474	474	6.58	1.16	1.11	9.83	72	—	71
31. Tonk	1122	49	1054	1103	10.99	3.02	2.86	9.88	119	8	70
32. Udaipur	2406	957	1449	2406	25.77	2.25	1.67	26.28	78	15	36
TOTAL	**42266**	**7208**	**33782**	**40990**	**447.8**	**82.26**	**71.97**	**451.63**	**4031**	**429**	**64**

Source: Disaster management and Relief Department

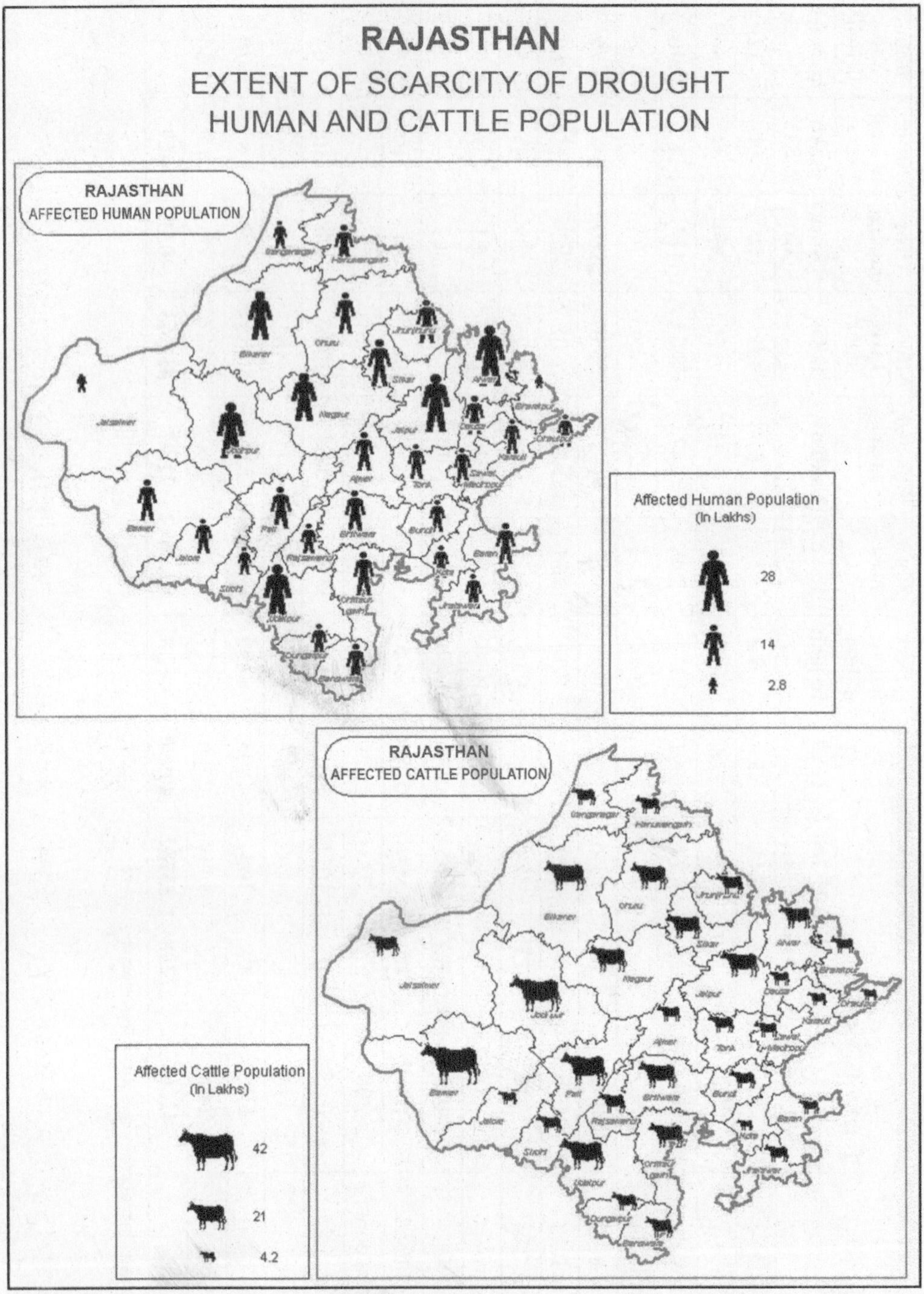

Source: Disaster Management and Relief Department

Fig. 24: Rajasthan Extent of Scarcity of Drought Human and Cattle Population

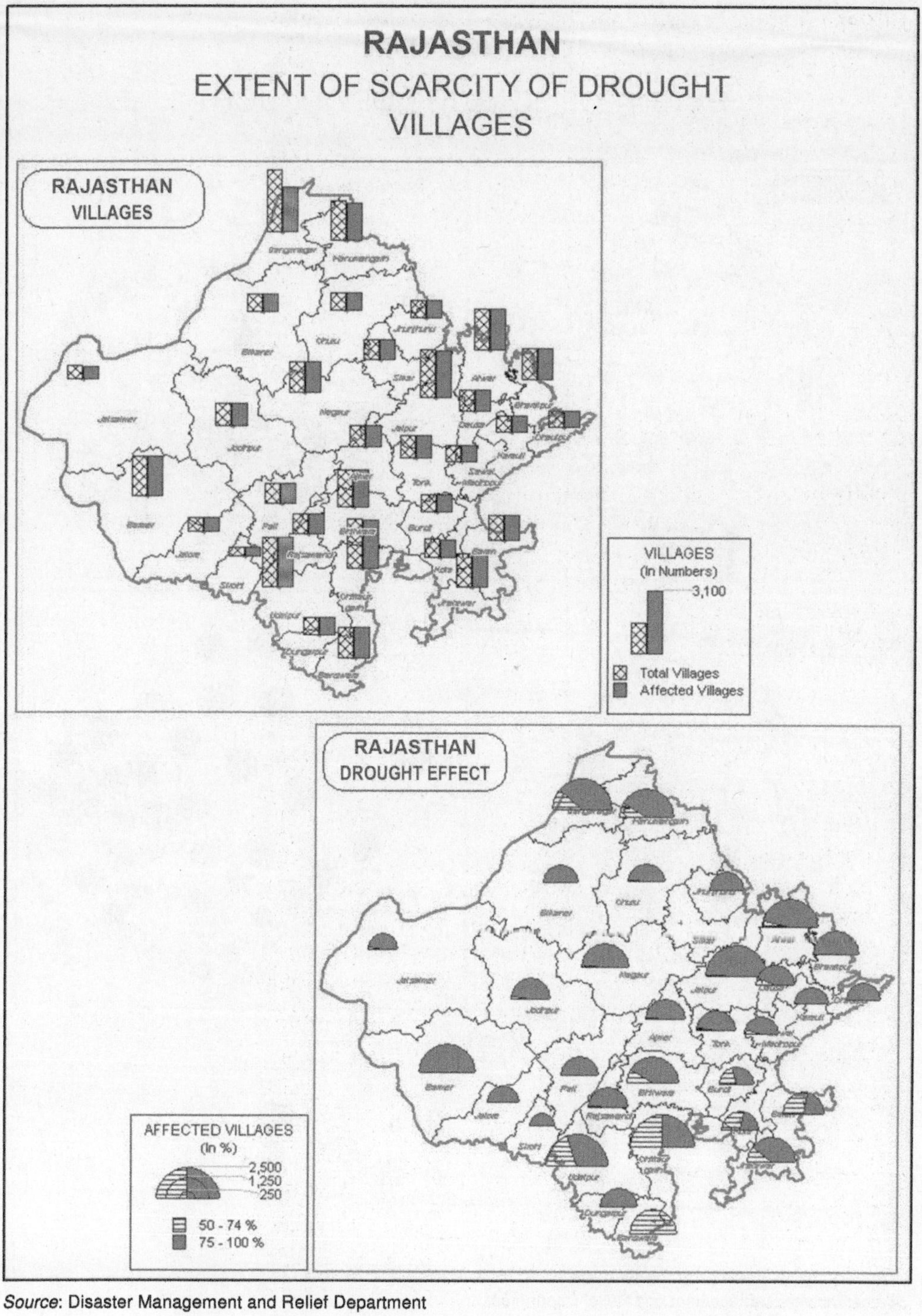

Source: Disaster Management and Relief Department

Fig. 25: Rajasthan Extent of Scarcity of Drought Villages

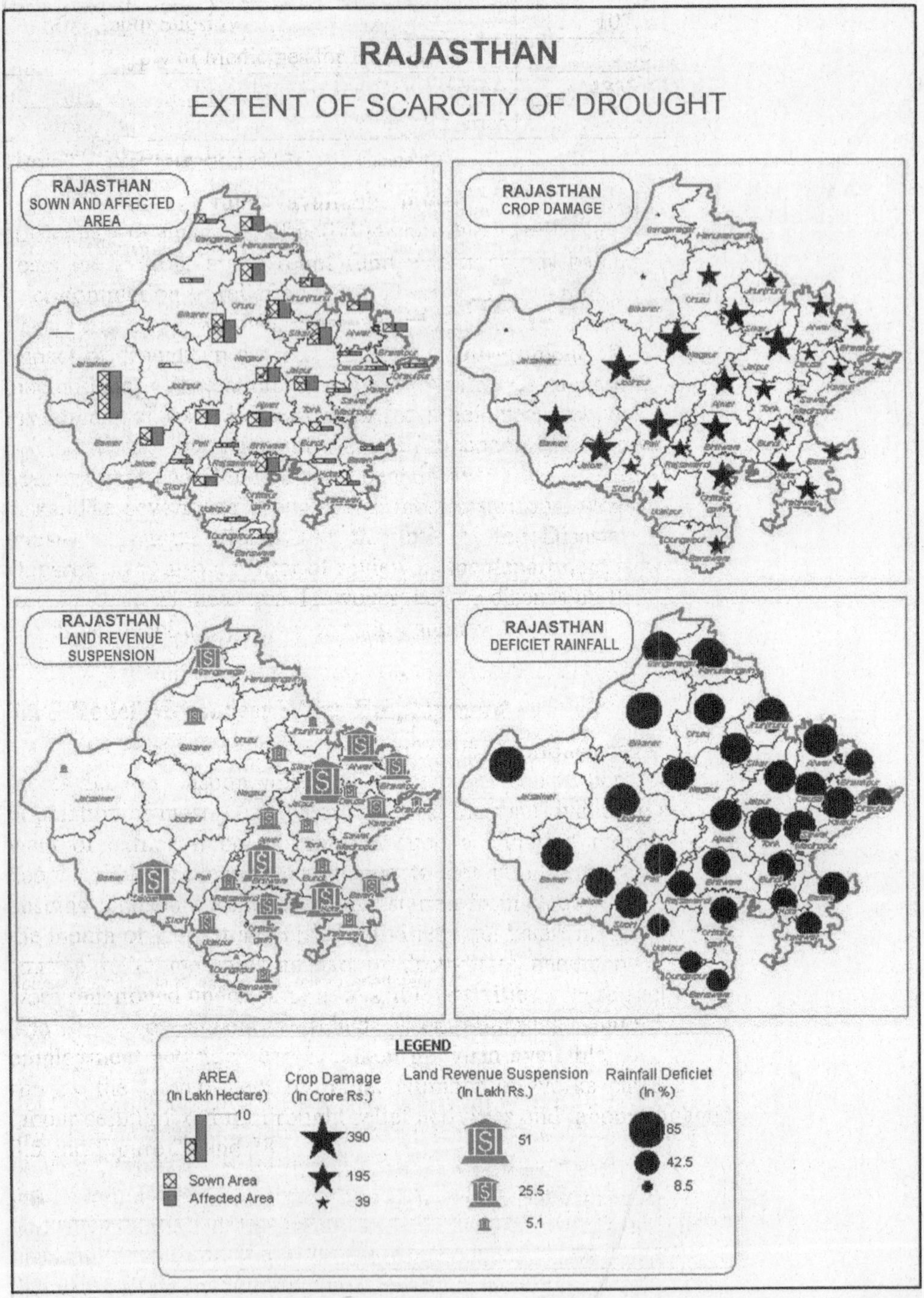

Source: Disaster Management and Relief Department

Fig. 26: Rajasthan Extent of Scarcity of Drought

Severity of drought can be imagined from the fact that 96.98 percent villages were affected with the drought of which 17.05 percent villages were affected with crop damage between 50-74 percent, while 79.93 percent villages had crop damage between 75 to 100 percent, where value of crop damage was Rs. 4031 crores and land revenue worth Rs. 429 lakhs was suspended. Out of crop sown area under kharif of 82.26 lakh hectares, crop in 71.97 lakh hectare area was severely affected. Annual rainfall deficit was (-)64 percent.

8.2.2 Financial Management of Drought 2002

The drought severity was highest of the century and state government submitted memorandum to the Government of India for providing financial assistance as the funds available in calamity relief fund were not sufficient to manage total relief measures. The central team visited state in response of the state government and submitted the report. The Amount received during the year 2002-03 and 2003-04 for relief activities undertaken from August 2002 to July, 2003 was the duration where next rainfall was due in July, 2003. The amount received for relief works and expenditure incurred on relief measures in given in Table 8.4.

Table 8.4: Receipt and Expenditure on Drought, 2002

Receipt under	Amount Received (Crore Rs.)	Balance (Crore Rs.)
Opening balance as on 1-4-2002		-150.30
A. AVAILABILLITY OF FUNDS		
CRF 2002-03 incl. State share	288.12	
CRF 2003-04 incl. State share	179.72	
NCCF 2002-03		
Goshala	11.66	
Input subsidy	155.68	
Fodder Transportation	7.40	
NCCF 2003-04		
Drought	622.44	
Goshala	14.47	
Further receipt for drought	99.83	
PM Relief Fund	4.58	
Total Receipt	**1383.90**	**1383.90**
B. ACTIVITY-WISE EXPENDITURE		
1. Relief Works	723.26	
2. Cattle Conservation		
(a) Fodder Transportation Subsidy	163.95	
(b) Cattle Feed Subsidy	19.08	
(c) Bull subsidy	0.74	
(d) Cattle Camp & Goshalas Subsidy	193.48	
(e) Medicines for animals	3.70	

Receipt under	Amount Received (Crore Rs.)	Balance (Crore Rs.)
3. Gratuitous Relief	22.61	
4. Drinking Water Transportation	67.12	
5. Supplementary Nutrition	30.00	
6. Input Subsidy	107.36	
7. Supply of Medicines for Human	1.33	
Expenditure	**1332.62**	**1332.62**
Balance		**- 99.02**

Source: Disaster management and Relief Department

Though the funds available under calamity relief fund and NCCF were sufficient to manage relief activities but due to relief activities carried out in earlier years the state calamity relief fund was in minus balance of 150.30 crores. Thus after completion of relief activities there was net minus balance of Rs. 99.02 crores. The drought of 2002 was most hazardous and the state was able to put the drastic impact of drought in complete state, the Government of India helped to the desired level. Though operation of relief activities is matter of serious discussion as investment of Rs. 1332.62 crores for relief measures was quite significant and its productive use is a matter of managerial capability of the state. The role of Disaster Management and Relief Department is also critical in fulfillment of the expected tasks. The severity of drought was most disastrous, where survival and human and livestock is first priority but the task of the Disaster Management and Relief Department is also a matter of review as the department is assigned one single target to manage relief measures. However, before discussing these aspects it is relevant to see the expenditure pattern of each category of the relief measures to arrive at the focused approach.

8.2.3 Relief Measures: Wage Employment

The cultivators of the state suffered a serious loss of crop worth Rs. 4414 crores due to monsoon failure, which is more than 50 percent crop loss. The affected population of mostly cultivators had lost their survival base and became helpless for want of extinct hope of crop. In such a situation cultivators are main affected people, who need some assistance to get purchasing power for the duration of ensuing monsoon. Relief measures started from October and crop loss was known in the month of August and relief activities could start in October, 2002. In addition to starting relief measures as part of drought management, wage oriented activities were sanctioned under other admissible activities, where such activities are possible. The state government searched potentialities in similar schemes where wage employment activities can be taken up form available funds of state and schemes run by the Government of India. Number of works carried out in each district, labour ceiling fixed for drought relief activities and labour engaged is given in Table 8.5 and Figure 27 and 28.

Table 8.5: Wage Employment Activities Carried Out in Districts

District	Wage Employment under Drought			Total Wage employment		
	No. of Works	Labour Ceiling	Labour Engaged	No. of Works	Labour Ceiling	Labour Engaged
1. Ajmer	2366	127700	124700	7978	163540	160550
2. Alwar	458	19900	18028	2136	41600	37181
3. Banswara	2613	90700	83530	4548	135900	104092
4. Baran	2090	99600	87536	5102	123200	114813
5. Barmer	2662	157200	155023	15455	208000	233613
6. Bharatpur	733	19300	21043	4895	36900	42802
7. Bhilwara	4203	145300	141053	9472	171300	172409
8. Bikaner	2802	133000	137214	16543	167400	215718
9. Bundi	1102	67700	80052	2754	84800	101795
10. Chittorgarh	4734	110500	107400	10429	149800	147800
11. Churu	1767	104200	85024	17902	145700	168337
12. Dausa	940	41200	41022	3239	54300	68445
13. Dholpur	145	15000	13900	1681	23900	25990
14. Dungarpur	3595	132100	150062	7034	1557700	191711
15. Ganganagar	323	25900	27607	1163	33400	33303
16. Hanumangarh	257	34300	16256	1542	55800	38223
17. Jaipur	3346	111000	98560	11544	155400	160784
18. Jaisalmer	1647	79500	78500	7686	101300	100298
19. Jalore	1813	87900	86506	2155	104700	92516
20. Jhalawar	1241	70000	51999	6880	95700	96608
21. Jhunjhunu	1195	62300	56589	2563	84000	63940
22. Jodhpur	2047	138700	138091	8876	171300	180247
23. Karauli	1648	84200	84895	4495	112600	110579
24. Kota	1189	38600	43240	2453	77600	79233
25. Nagaur	1864	109800	106111	9485	153600	157872
26. Pali	1739	139400	150075	2934	172400	168105
27. Rajsamand	2552	84400	80290	6692	105600	106790
28. S.Madhopur	1078	97800	97800	3921	121000	121000
29. Sikar	970	60000	47123	5499	90200	94227
30. Sirohi	1512	80800	77554	4956	96500	100134
31. Tonk	1183	106000	100626	1947	121880	108299
32. Udaipur	4098	194600	160251	8456	218420	199066
TOTAL	**59912**	**2869500**	**2747660**	**202418**	**3735440**	**3796480**

Source: Disaster management and Relief Department

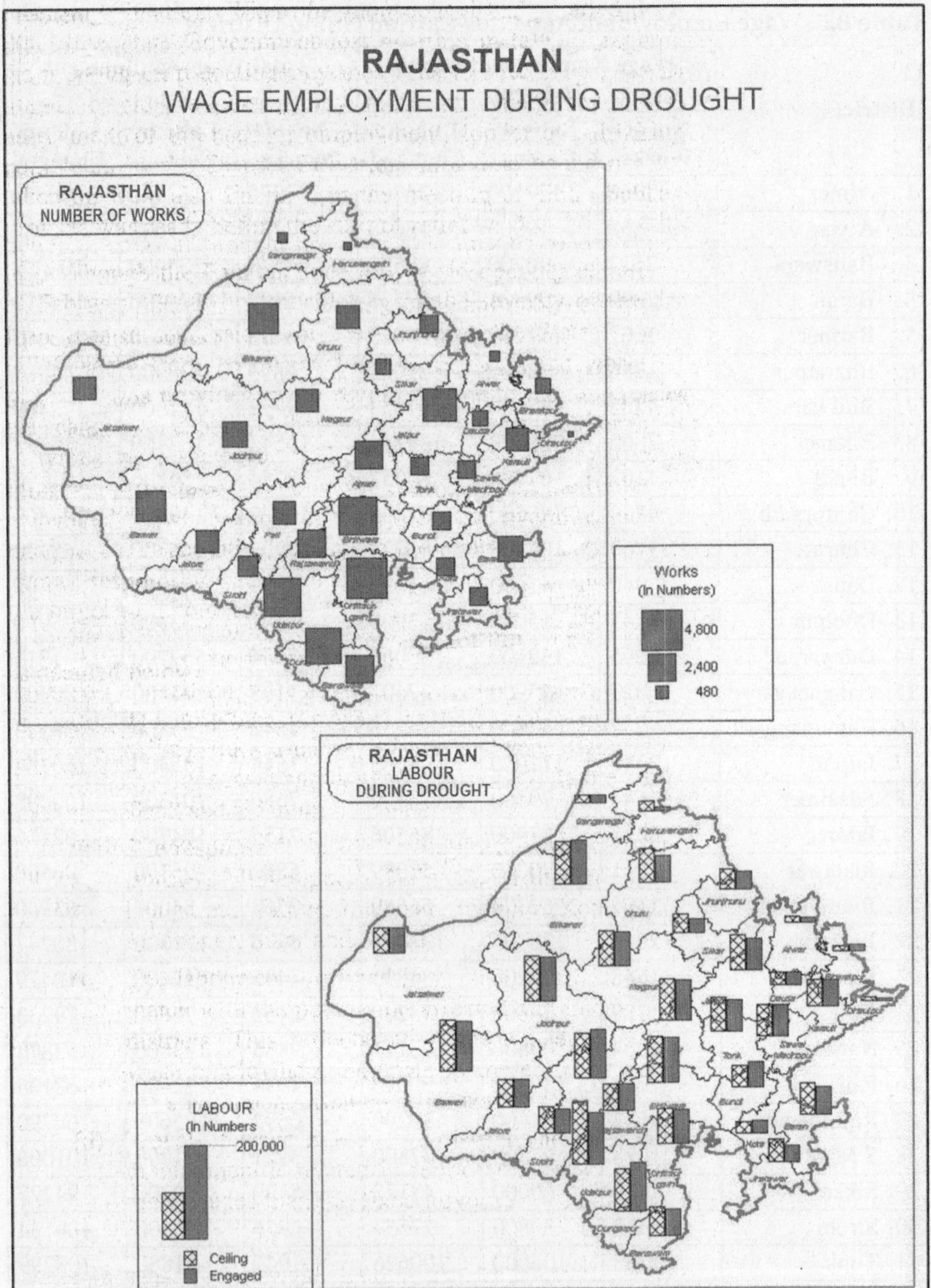

Source: Disaster Management and Relief Department

Fig. 27: Rajasthan Wage Employment During Drought

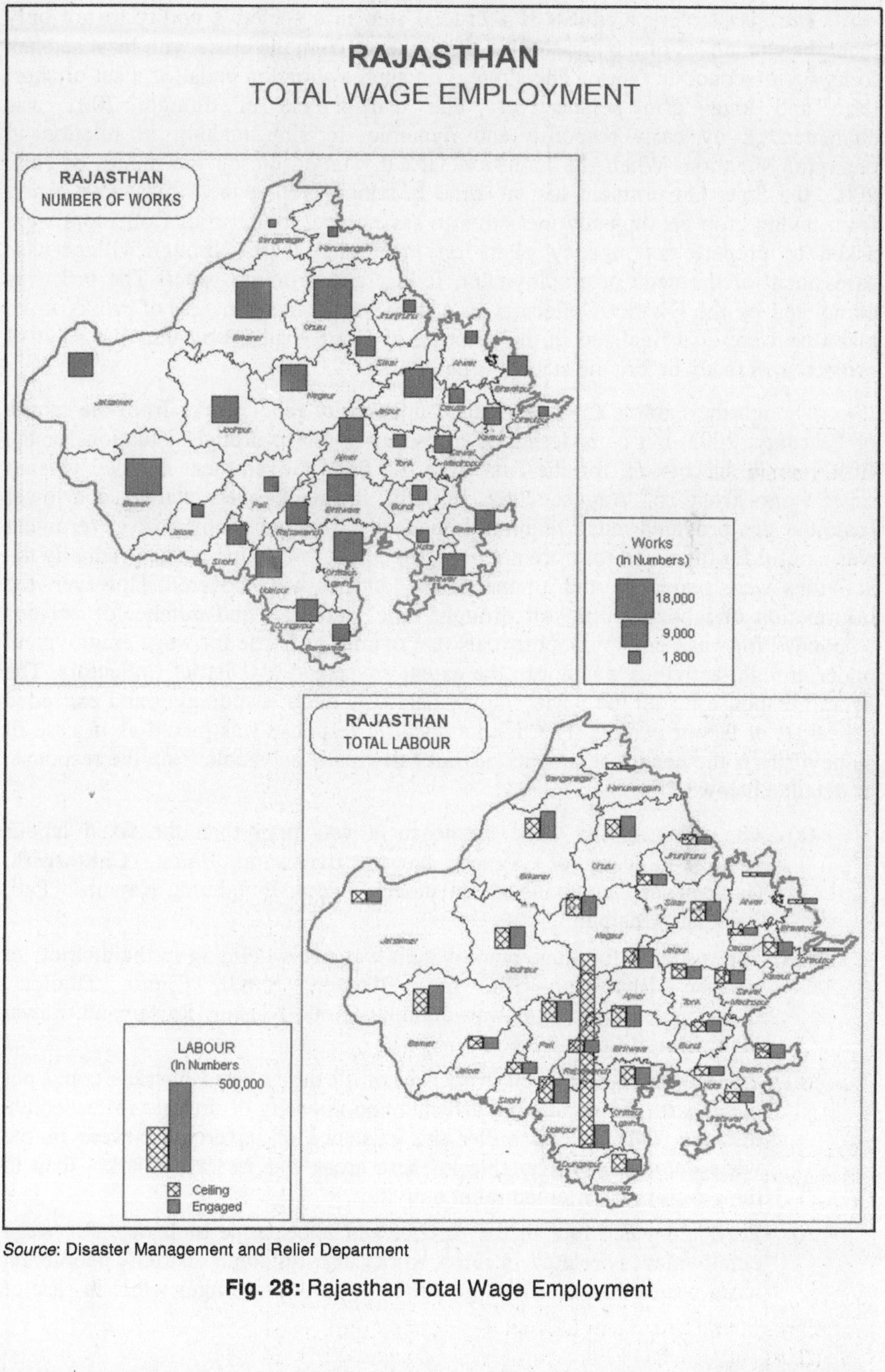

Source: Disaster Management and Relief Department

Fig. 28: Rajasthan Total Wage Employment

The disaster preparedness is a crucial step in a society's ability to not only meet the challenges of the disaster but also emerge from the crisis with least setback to its socio-economic fabric. The strategy in such a situation calls for a set of short term and long term perspectives. The management of drought 2002 was characterized by early response and dynamic decision making in relation to emerging situation. When the monsoon virtually failed during the month of July, 2002, the State Government lost no time in taking preparedness measures. Apart from taking crop contingency measures to save crops, the District Collectors were asked to prepare contingency plans on emergency basis through village-wise assessment of the need of employment, fodder and drinking water. The task was completed by the District Collectors in a time bound manner. List of works to be taken up were also finalized in the meeting of Gram Sabhas and thus the shelf of projects was ready before the start of relief works.

Though the Famine Code envisaged opening of relief works from the month of February, 2003, but considering of the severity of the drought situation for the fifth year in succession, for the first time the State Government decided to start relief works from 16th August, 2002. Initially, the works were started at a lower scale but this provided much required hope to the affected people that government was careful for them and they are not going to suffer for famine code. Gradually the activities were scaled up and all the needy families were covered. However, the information of labour ceiling for drought relief activities and number of persons responded for wage employment reveals that overall response for wage employment under drought activities was not to the extent envisaged by District Collectors. The overall response for all the wage employment was quite encouraging and exceeded the target of labour ceiling. The district specific response was mixed as in case of some districts the need assessment was not fully made as visible from the response, as detailed below:

(a) The response for wage employment was poor than the fixed labour ceiling in case of Ajmer, Alwar, Banswara, Baran, Chittorgarh, Ganganagar, hanumangarh, Jaisalmer, Jalore, Jhunjhunu, Karauli, . Pali, Tonk and Udaipur.

(b) The response for wage employment was overwhelming in the districts of Barmer, Bharatpur, Bhilwara, Bikaner, Bundi, Churu, Dholpur, Dungarpur, Jaipur, Jhalawar. Jodhpur, Kota, Nagaur, Rajsamand, Sawai Madhopur, Sikar and Sirohi.

(c) The labour ceiling fixed in respect of all the affected districts could not match with the population affected and intensity of drought in respective districts. This reflects either the existence of poverty is severe or the wage employment available in these areas was lesser in market than in the government oriented relief activities.

(d) Prevalent wage rate in the district and appearance of people for wage employment is related in relief works also. In some districts, people get more wages than the fixed government minimum wages while in case of other districts people turn up for relief works as prevalent wage rate in other sectors is lesser in those districts.

(e) In some cases, where turn out of people for wage employment activities, is governed by various factors, including political effects to get higher labour ceiling, which is not in consonance with the drought impact. The higher turn up may also possible in the districts where misappropriation of funds remain higher.

(f) This is general practice that people of most severe drought affected areas turn up for wage employment as their crops are damaged to maximum extent, but such conditions are not reflected in determining the labour ceiling. This reveals that there are some other factors which play their role in fixing labour ceiling.

(g) Misappropriation of funds in relief activities are higher in comparison to normal wage oriented activities. Various schemes like, Sampoorna Grameen Swarojgar Yojna, DDP, DPAP are being implemented on regular basis but turn out of people for wage employment remains lower than the envisaged number and funds received from the Government of India remain unutilized every year. Audit objections in relief works are highest as visible from past and year 2002 relief analysis.

(h) State Government complains that the Government of India does not allow material cost in case of drought relief works and in such a situation, state Government is compelled to allow kachha works for this helplessness.

(i) This is also true that in drought relief measures, the number of incomplete activities remain higher, which are either left out unattended when the relief work is over. In such a situation, the position of uncompleted works remains problem for the state government as their remains no resources to complete such tasks from regular development works.

(j) Wage employment measures are carried out with imaginary works as where about of most of the activities carried out during famine relief are never visible in the field. There are most strict instruction to District Collectors to complete relief works on most priority basis but such guidelines are carried out without consulting the admissibility of such tasks, which has no relevance.

8.2.4 Cattle Conservation Measures

In the livestock sector, measure problem is faced related to cattle, which are usually not economic holding but people rear for religious and ethnic reasons. Small livestock like sheep and goat are generally migratory nature and manage their food by grazing of straw over the earth surface. Buffaloes are feeded in view of their livelihood support. Cattle are generally left unattended in case of drought conditions. There are some goshalas working in the state to protect cattle. Male animals used in cultivation are became useless due to mechanized farming and their number is decreasing continuously. During drought years, government provides fodder to cattle for survival. During drought of 2002, condition of cattle became miserable as there was no fodder to feed. This is regular programme of the state government to take care of cattle during drought year and during the drought of 2002, significant efforts were made through various modes. Measures adopted during the drought 2002 are given in Table 8.6.

Availability of food is possible from the stock preserved by the Government of India and food grain is allotted to drought affected states as per assessed demand. In case of fodder, it is only possible from the neighbouring states where fodder remains surplus than their own domestic requirement. The state is required to bear cost of fodder and transportation charges for carrying to deficient areas. This problem is generally felt in desert districts except Ganganagar and Hanumangarh as these districts also have surplus fodder to meet district requirement. During drought 2002, fodder worth 19.40 lakh MT was provided to deficient areas except Kota district. Fodder was distributed to needy people through 5448 depots. For this purpose 5529 depots were sanctioned but 81 depots could not be opened for administrative reasons (Fig. 29-33).

The state government had also thought for establishment of fodder bank in deficient areas as drought proofing measures as fodder shortage is usually realized in the state and crop failure is recurrent feature. Cattle camps were also organized and against sanctioned 2173 cattle camps only 2029 camps were opened. The theme for establishment of cattle camp was necessitated in view of the fact that poor people are not able to manage their cattle during severe drought conditions and left those to survive at their own. In most of the cases, such cattle die as there remains no vegetative cover over the earth surface during drought conditions and even dry straw, bushes etc are used by sheep and goat unable to be migrated. Cattle camps are established tu survive such cattle and people are required to carry those to camps and take back after rains. 5.78 lakh cattle were managed in these camps. These camps provide fodder and water to needy cattle and try to survive them in most severe conditions.

Goshalas are run by private organizations like trusts, which receive funds for maintenance of cattle for protection of cattle breed. Earlier these organizations were engaged in breed improvement but now, their role has been restricted to keep left out cattle wealth. There were 804 goshalas in 30 districts aimed to maintain the cattle wealth to survive wandering cattle and pass own to goshalas possessing more than 3 lakh cattle wealth. Maintenance of bulls is also necessary for breed improvement. During normal conditions bulls are left free to graze in the surrounding areas of village and meet food requirement by grazing in open lands. During drought conditions 4107 bulls in five districts were provided fodder to survive.

Table 8.6: Cattle Conservation Measures During Drought 2002

District	Fodder Arrangement		Cattle Camps		Goshalas		Bull benefited	Cattle Feed subsidy	
	No. of depots	Quantity distributed (MT)	No.	Cattle benefited	No.	Cattle benefited	No.	Cattle Ceiling	Cattle benefited
1	2	3	4	5	6	7	8	9	10
1. Ajmer	323	130741	9	4200	16	19521	15	51000	43092
2. Alwar	26	2631	—	—	8	3619	—	—	—
3. Banswara	29	813	—	—	3	470	—	—	—
4. Baran	46	821	26	13700	6	2232	—	—	—
5. Barmer	478	440946	427	170220	18	10430	—	54000	68740
6. Bharatpur	21	1107	6	4942	3	964	—	—	—
7. Bhilwara	559	198313	—	—	16	5077	—	15000	4757
8. Bikaner	396	213786	162	101754	31	18775	—	56000	56000
9. Bundi	12	96	1	300	11	4400	—	—	—
10. Chittorgarh	110	7303	2	550	11	1932	—	5000	8278
11. Churu	256	223093	69	16002	28	8655	898	30000	30000
12. Dausa	30	4450	11	2700	—	—	—	—	—
13. Dholpur	11	200	5	2500	3	1050	—	—	—
14. Dungarpur	182	16566	3	325	—	—	—	—	—
15. Ganganagar	2	8	35	8195	18	9951	—	—	—
16. Hanumangarh	25	23773	42	7300	38	16266	—	—	—
17. Jaipur	165	29101	29	4716	33	6286	—	10000	9546
18. Jaisalmer	324	134431	594	120000	35	28640	—	45000	45000
19. Jalore	132	47694	17	3820	38	43297	—	15000	16611

District	Fodder Arrangement		Cattle Camps		Goshalas		Bull benefited	Cattle Feed subsidy	
	No. of depots	Quantity distributed (MT)	No.	Cattle benefited	No.	Cattle benefited	No.	Cattle Ceiling	Cattle benefited
1	2	3	4	5	6	7	8	9	10
20. Jhalawar	27	2125	2	595	7	2245	—	—	—
21. Jhunjhunu	124	35662	14	4866	24	6535	—	—	—
22. Jodhpur	555	145727	128	29829	181	47438	1232	65200	58320
23. Karauli	122	1451	11	2870	4	1292	—	26000	1873
24. Kota	—	—	—	—	8	1843	—	—	—
25. Nagaur	165	35707	144	22600	121	21783	1174	10000	10000
26. Pali	328	83508	113	26964	57	18624	—	35000	31985
27. Rajsamand	176	57216	5	2030	5	2435	—	10000	6300
28. S. Madhopur	152	9600	18	4800	6	1550	—	25000	50000
29. Siikar	22	7373	54	7980	53	11068	788	10000	2396
30. Sirohi	110	19435	8	2299	9	8379	—	13000	23845
31. Tonk	183	33748	10	1609	8	2994	—	—	—
32. Udaipur	357	31536	84	10516	5	784	—	20000	20000
Total	**5448**	**1939863**	**2029**	**578182**	**804**	**308535**	**4107**	**495200**	**486743**

Source: Disaster management and Relief Department

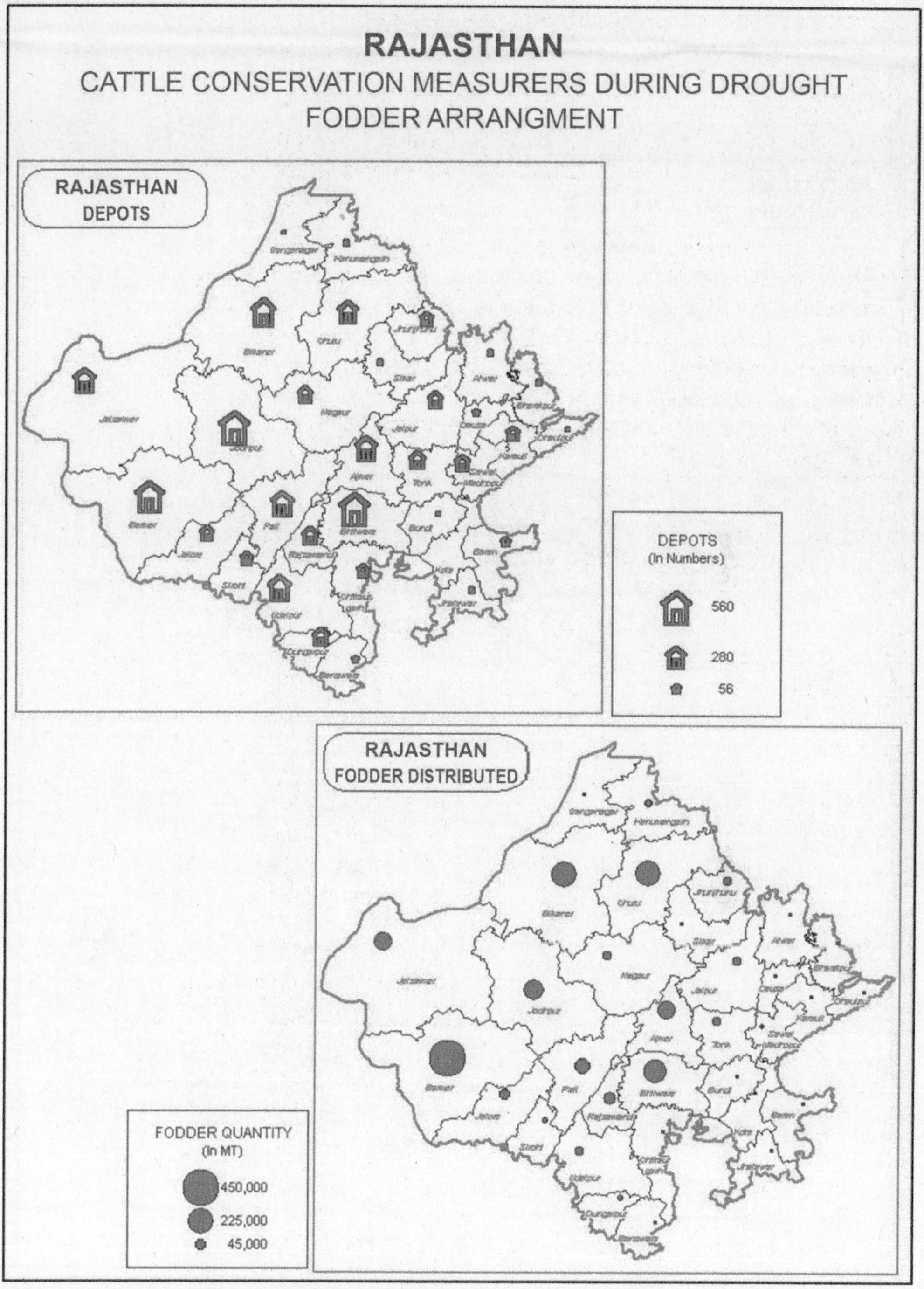

Source: Disaster Management and Relief Department

Fig. 29: Rajasthan Cattle Conservation Measures During Drought Fodder Arrangment

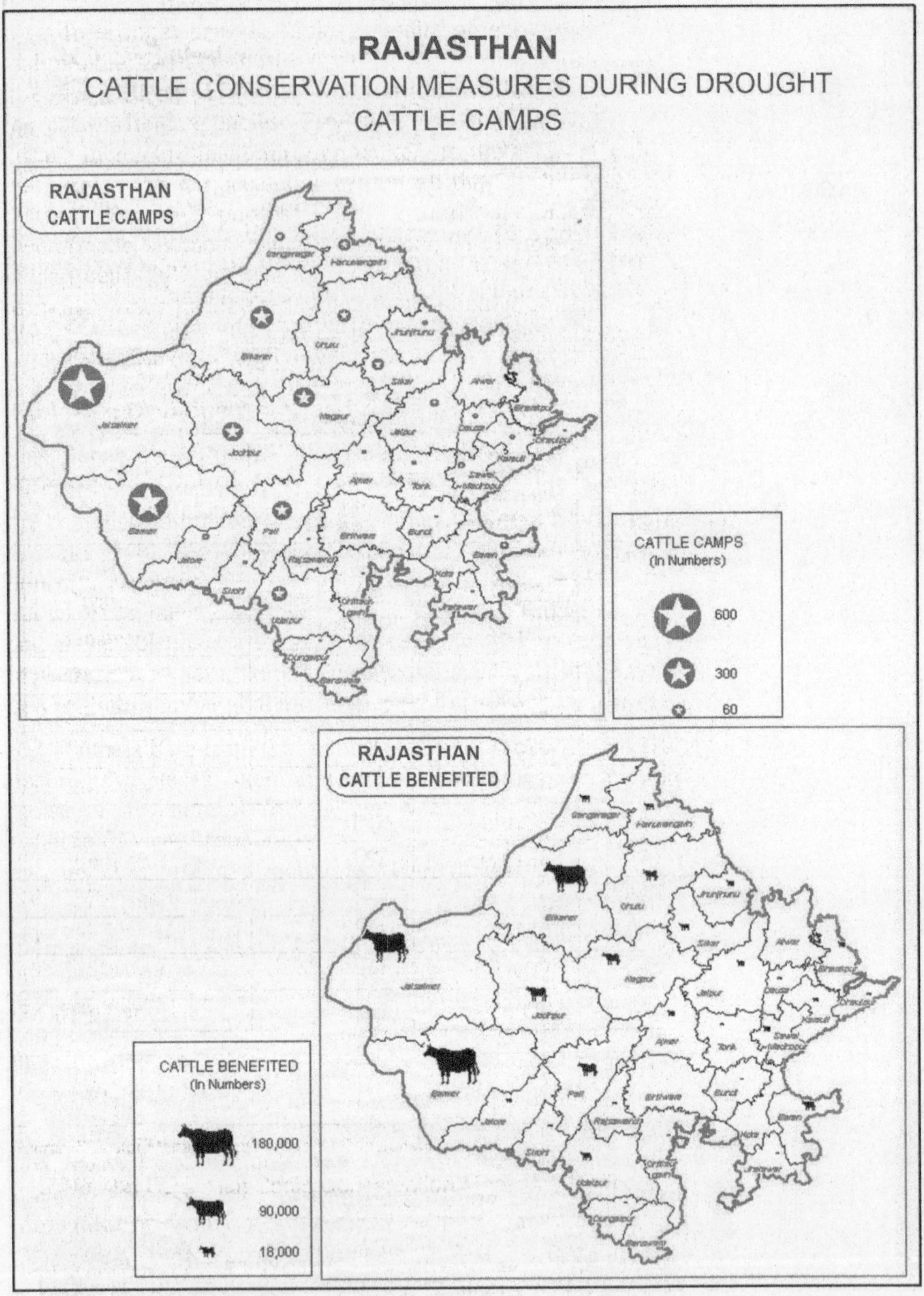

Source: Disaster Management and Relief Department

Fig. 30: Rajasthan Cattle Conservation Measures During Drought Cattle Camps

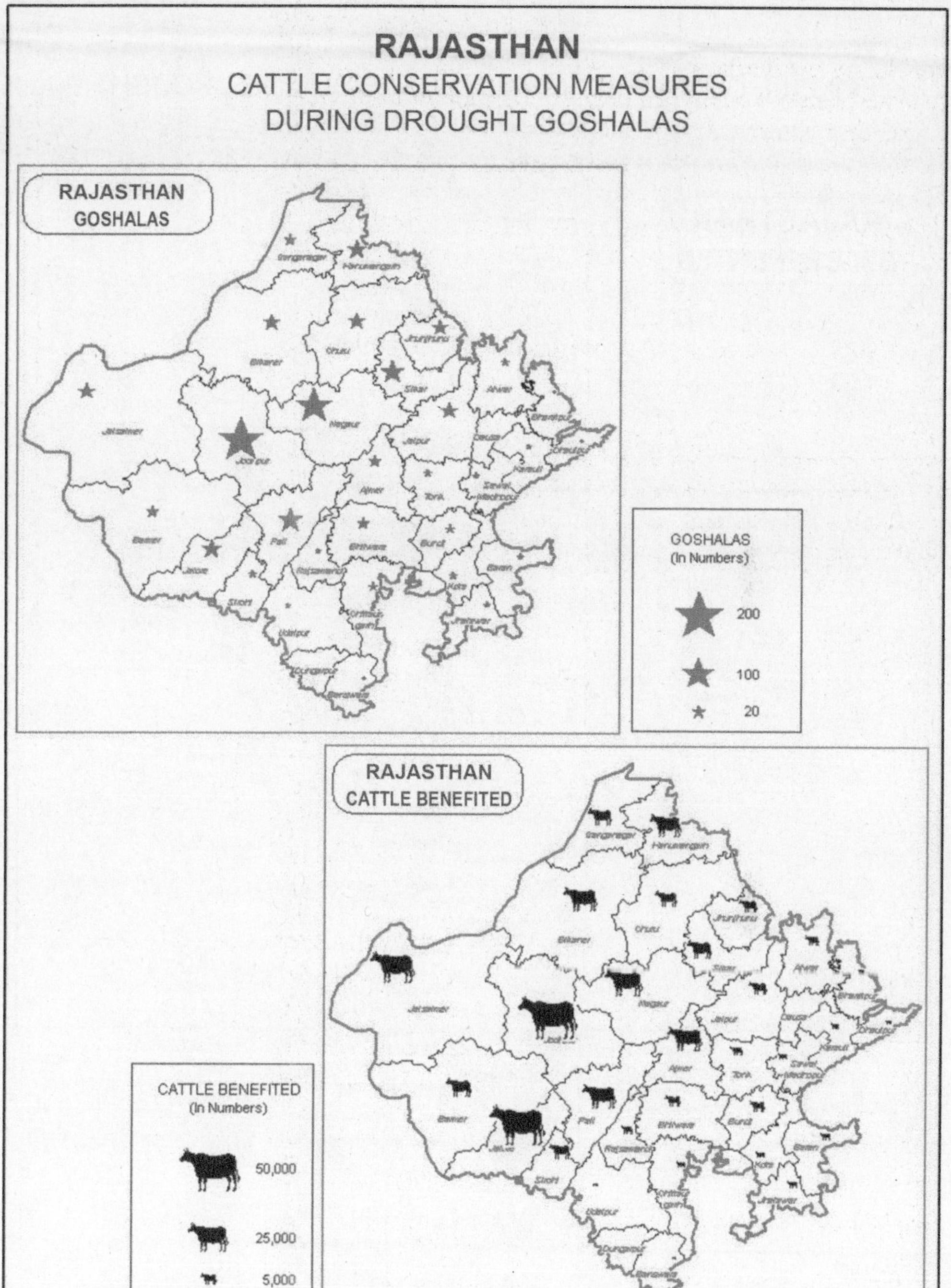

Source: Disaster Management and Relief Department

Fig. 31: Rajasthan Cattle Conservation Measures During Drought Goshalas

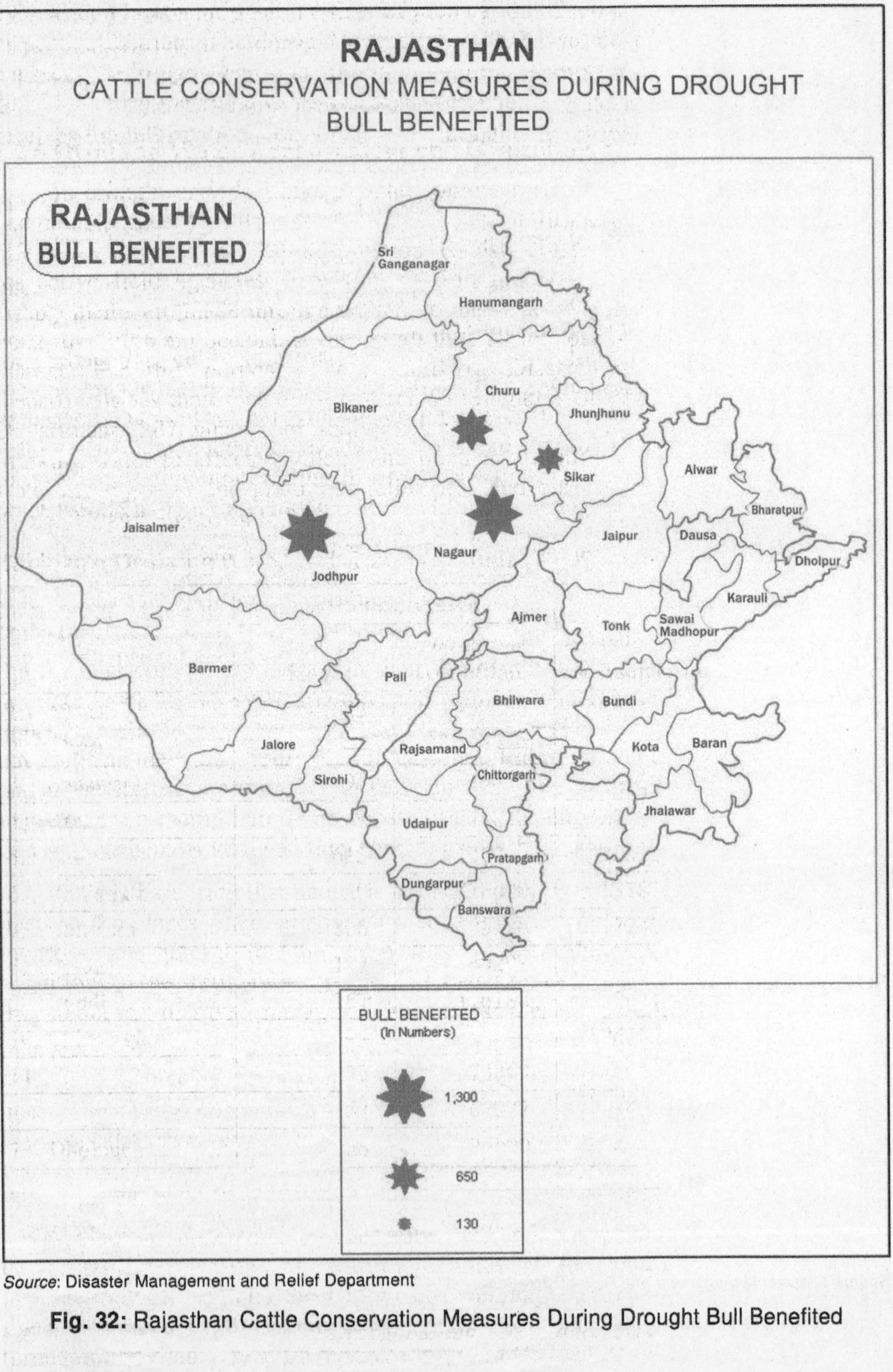

Source: Disaster Management and Relief Department

Fig. 32: Rajasthan Cattle Conservation Measures During Drought Bull Benefited

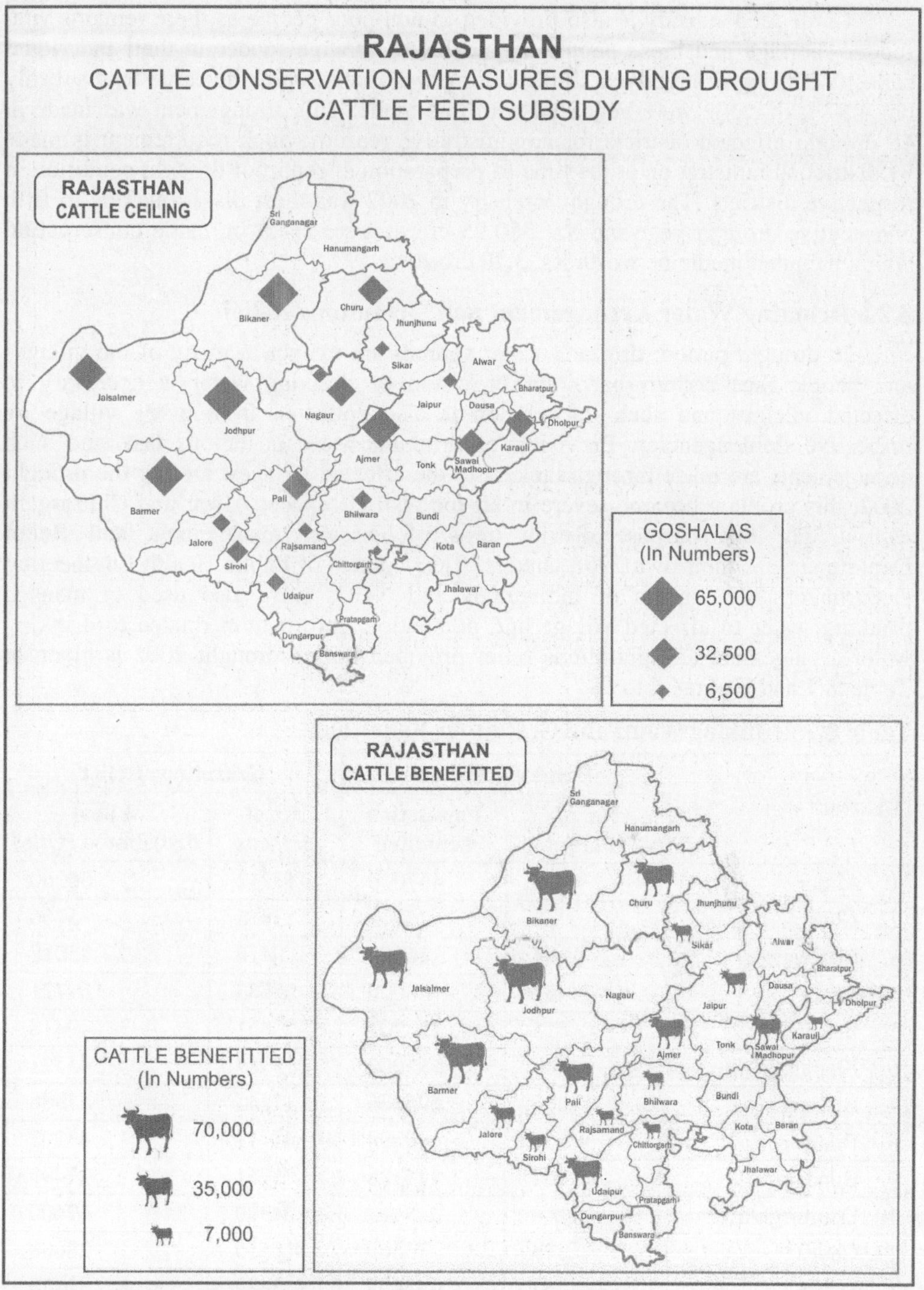

Source: Disaster Management and Relief Department

Fig. 33: Rajasthan Cattle Conservation Measures During Drought Cattle Feed Subsidy

Cattle feed subsidy is also provided to non-poor people as there remains vital fodder shortage and these people are unable to arrange fodder in their individual capacity. During drought year 2002, ceiling was fixed to provide cattle feed subsidy but 4.87 lakh cattle were benefited with this mode. This arrangement was made in 18 drought affected districts for administrative reasons. Such requirement is made by district administration at the time of preparation of report of drought condition in respective districts. The drought intensity in 2002 was most disastrous due to fifth consecutive drought year and Rs. 380.95 crores were spent on cattle conservation which included medicine worth Rs. 3.70 crores.

8.2.6 Drinking Water Arrangement and Gratuitous Relief

In drought period, drinking water sources are extinct in many of the villages and people face serious hardship. Provision of drinking water is necessary in affected villages and such information is also collected from every village by respective field agencies. Provision of drinking water is tedious task and such arrangements are made through tankers to the affected villages. During the drought 2002, this problem became severe in all the districts except Alwar and Chittorgarh district. For this purpose district units of Disaster management and Relief Department has their own 2998 tankers, 9133 tankers of Public Health Engineering Department, 387 tankers of individuals and NGOs ware also used to manage drinking water to affected village and population. Information related to drinking water arrangement and gratuitous relief provided during drought 2002 is given in Table. 8.7 and Figure 34 to 38.

Table 8.7: Drinking Water and Gratuitous Relief 2002

District	Drinking Water		Gratuitous Relief	
	No. of villages	Population benefited	No. of persons	Wheat distributed (Qtls)
1. Ajmer	622	467000	13212	66060
2. Alwar	—	—	3875	5766
3. Banswara	94	16450	8416	25018
4. Baran	67	48000	25633	69721
5. Barmer	615	138704	20851	49418
6. Bharatpur	40	15000	13787	43521
7. Bhilwara	806	695126	17245	57908
8. Bikaner	157	49298	45411	43207
9. Bundi	167	116188	5751	11490
10. Chitttorgarh	—	—	24300	47400
11. Churu	73	73369	10555	28008
12. Dausa	57	67077	3489	2478
13. Dholpur	38	24000	4573	2287

District	Drinking Water		Gratuitous Relief	
	No. of villages	Population benefited	No. of persons	Wheat distributed (Qtls)
14. Dungarpur	95	137200	9053	4537
15. Ganganagar	11	13721	11498	3558
16. Hanumangarh	15	25964	1989	19016
17. Jaipur	588	595000	19974	40952
18. Jaisalmer	491	46167	26091	15474
19. Jalore	500	178759	10166	33939
20. Jhalawar	65	226500	11689	49716
21. Jhunjhunu	27	14799	3463	8586
22. Jodhpur	1360	345674	24485	68005
23. Karauli	352	119596	7961	1098
24. Kota	19	23600	59739	34859
25. Nagaur	244	81600	18548	54480
26. Pali	310	258424	17070	27829
27. Rajsamand	533	325502	9931	31752
28. Sawai Madhopur	71	162667	10823	37524
29. Sikar	143	75590	8876	23053
30. Sirohi	199	172065	5676	10151
31. Tonk	109	80200	8619	21535
32. Udaipur	800	722091	50983	4290
Total	**8668**	**5315331**	**513732**	**942636**

Source: Disaster management and Relief Department

Analysis of drinking water arrangement is affected villages reveals that about 20 percent villages were affected with the drinking water problem and about 12 percent of rural population was required to provide drinking water due to dry sources. The affected villages are provided drinking water through wells and hand pumps. The analysis reveals that most severe problem was faced by the people of Ajmer, Barmer, Bhilwara, Bundi, Dungarpur, Jaipur, Jalore, Jhalawar, Jodhpur, Karauli, Pali, Rajsamand, Sirohi and Udaipur districts. This is in view of the fact that after provision of drinking water arrangements by the government, people neglected their own water harvesting practices at individual and community levels. Surprisingly out of fourteen affected districts, only four are desert districts. Thus from rainfall deficiency and drinking water problem, there is no match in the district arrangements. Actually, such problems are the regular feature of the departments responsible for drinking water arrangements.

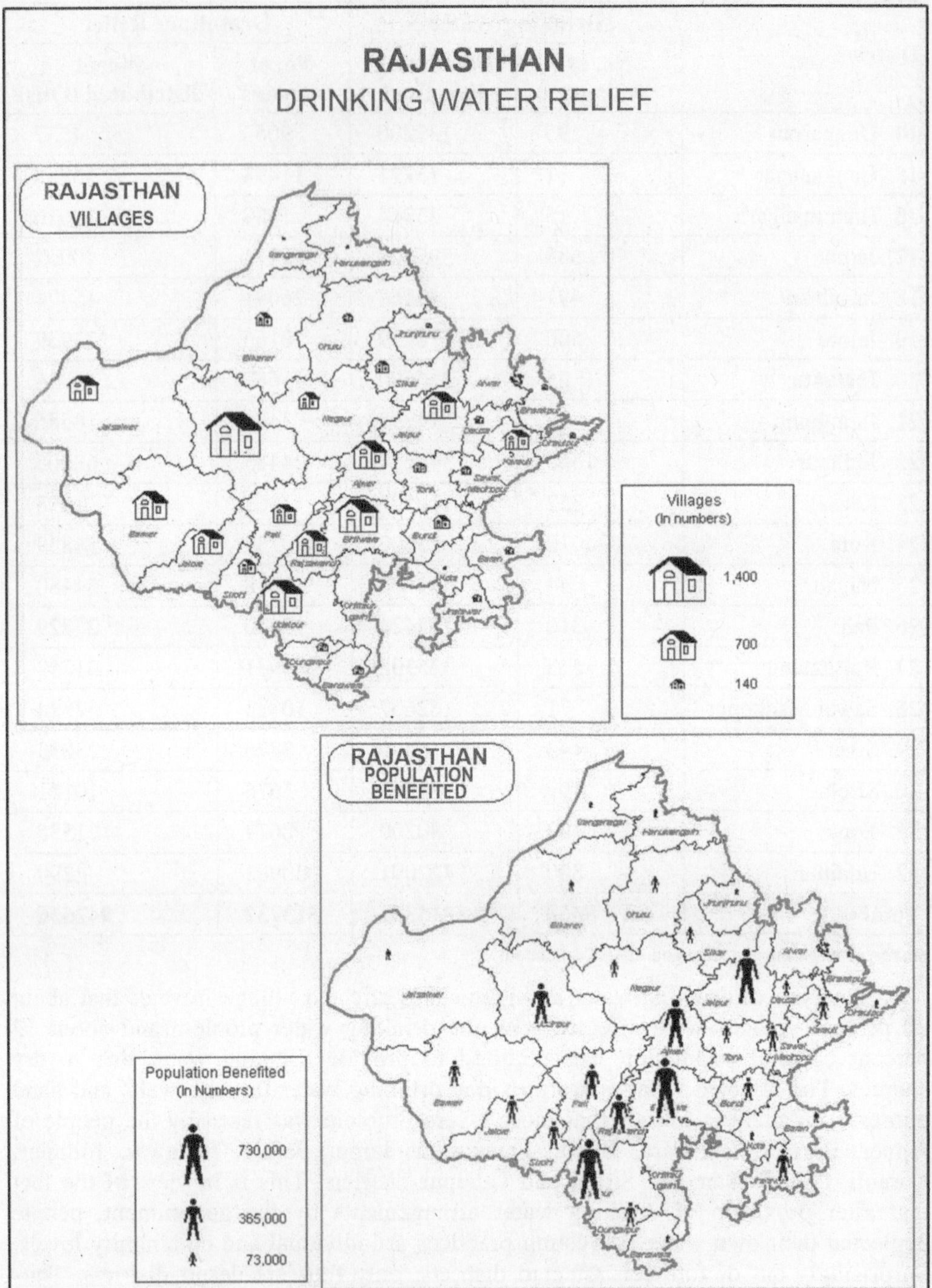

Source: Disaster Management and Relief Department

Fig. 34: Rajasthan Drinking Water Relief

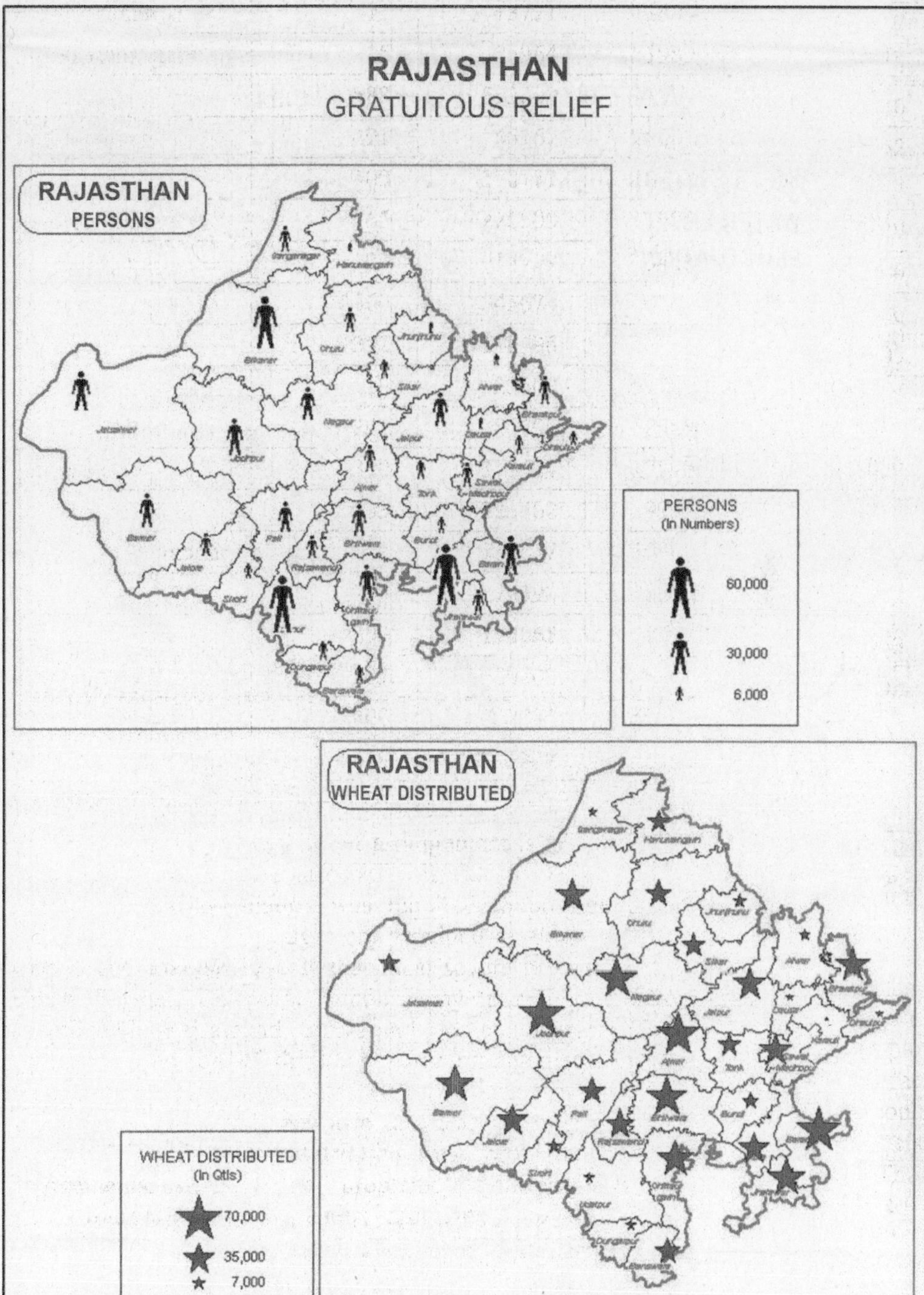

Source: Disaster Management and Relief Department

Fig. 35: Rajasthan Gratuitous Relief

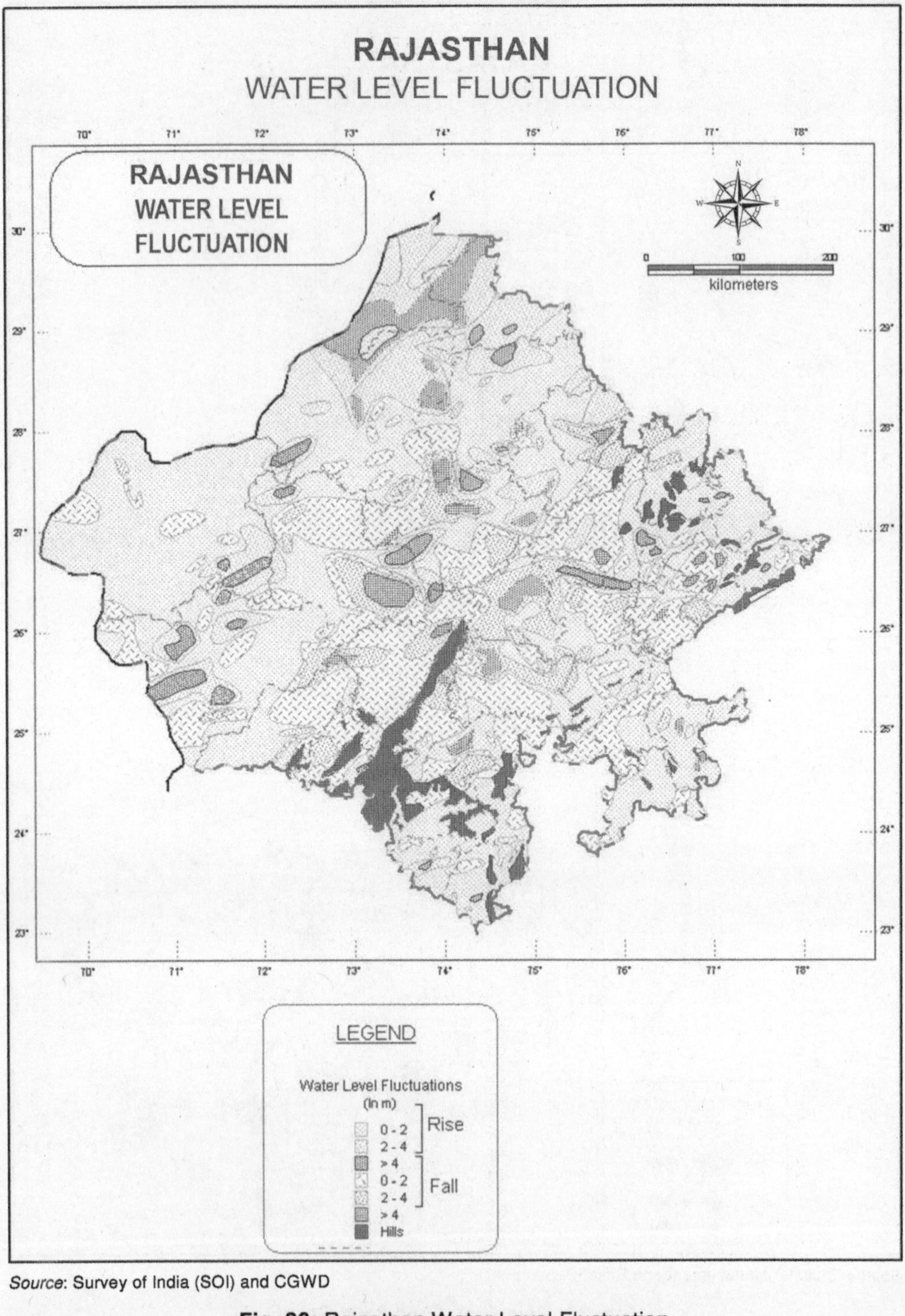

Source: Survey of India (SOI) and CGWD

Fig. 36: Rajasthan Water Level Fluctuation

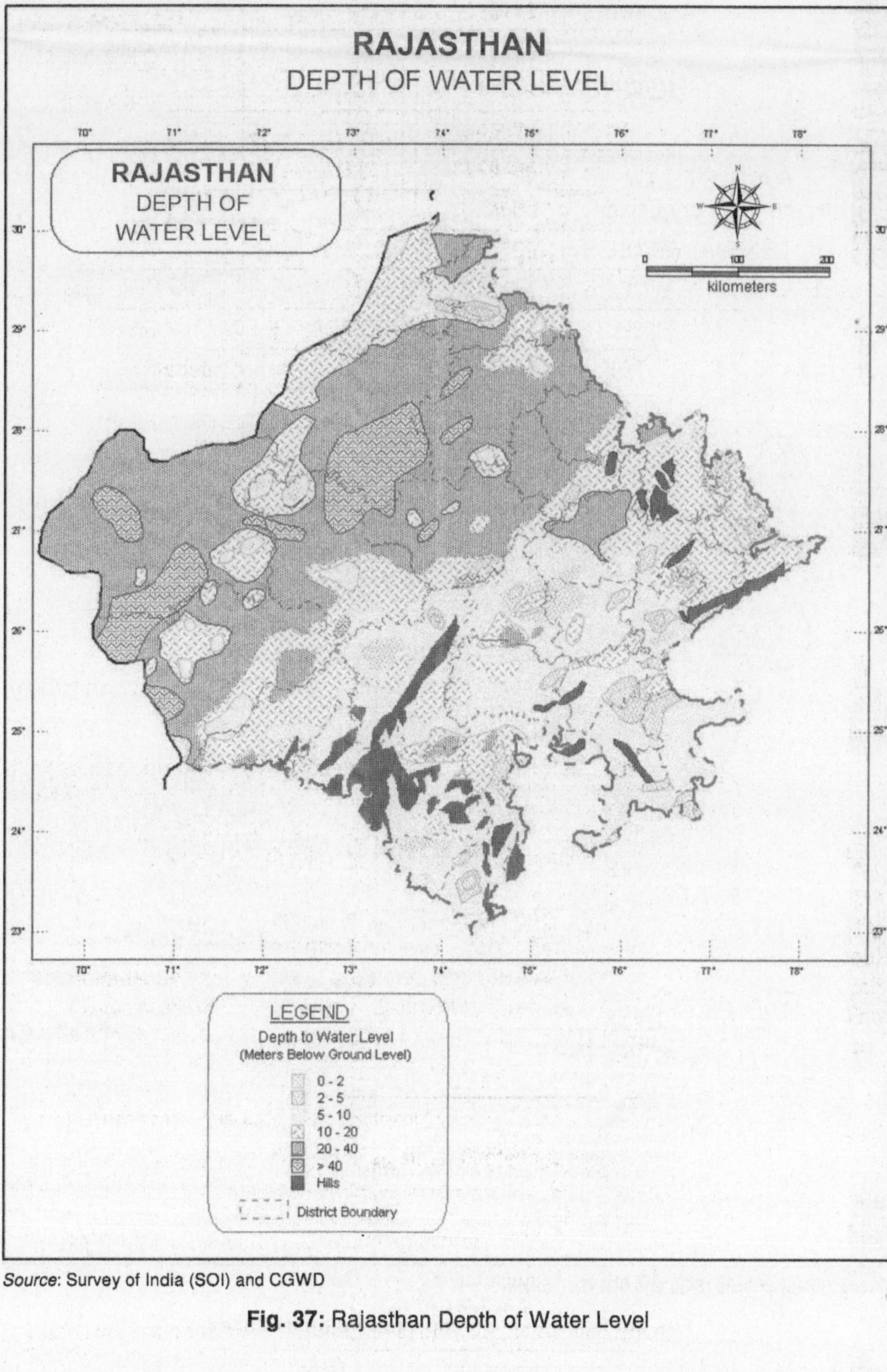

Source: Survey of India (SOI) and CGWD

Fig. 37: Rajasthan Depth of Water Level

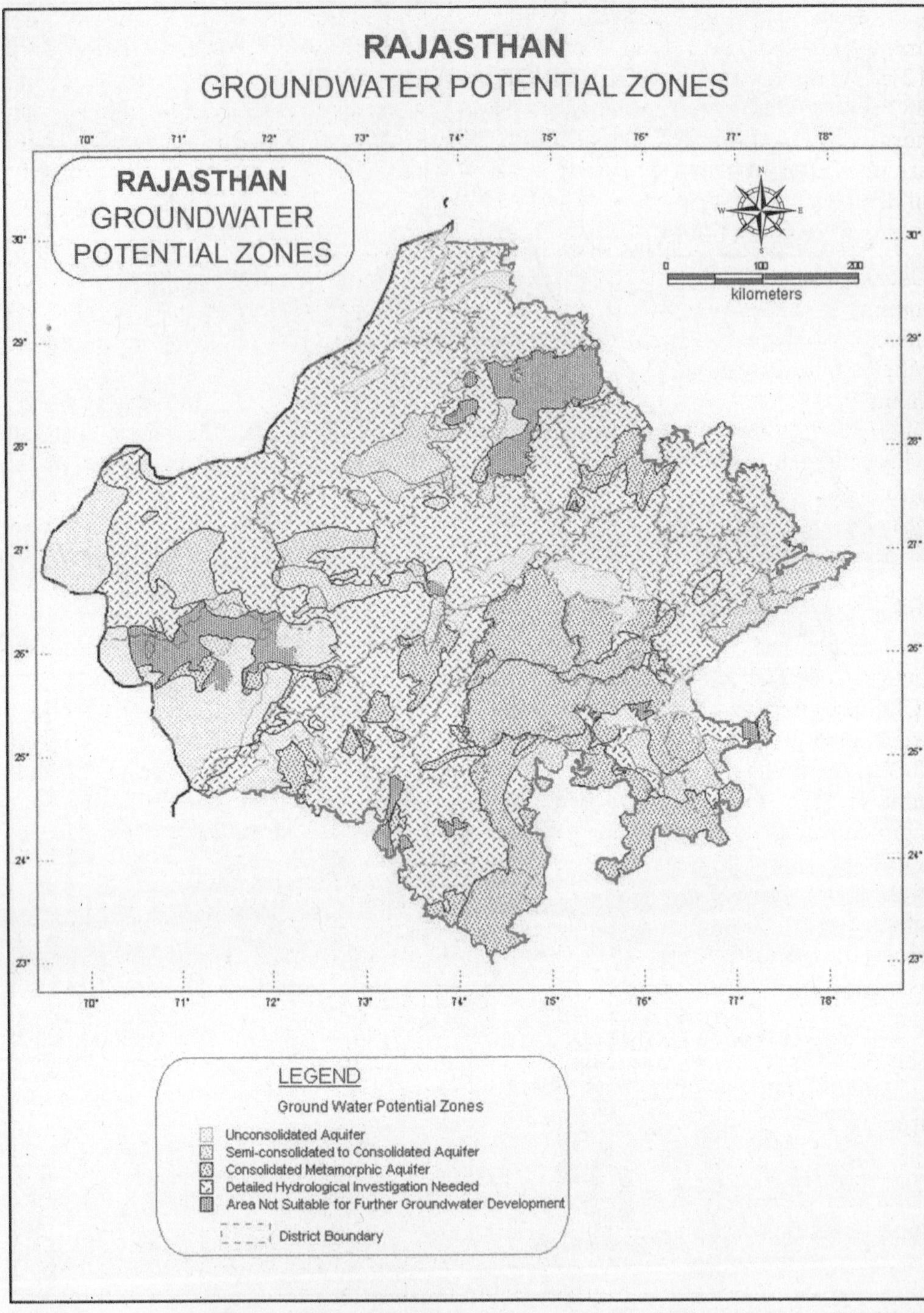

Source: Survey of India (SOI) and SGWD, Jodhpur

Fig. 38: Rajasthan Groundwater Potential Zones

Usually, during drought period, administration is required to manage arrangements as per drought severity and drinking water problems. There were 12131 water tankers of Relief and PHED for providing water to affected areas, but 387 private tankers were also associated in various districts. In addition there are no norms for use of the services of tankers and population benefited. These arrangements are usually made to avoid any casualty of life of the people for want of drinking water.

Gratuitous relief is provided to old and infirm people mainly to save their life. Usually one such person need maximum 1.50 quintal wheat for the duration of ten months of drought duration from August to June. The district analysis reveals that there was unequal distribution of wheat to such persons in the name of gratuitous relief. Surprisingly wheat distributed in Ajmer district of 5 quintals and Hanumangarh at the rate of more than 9 quintal per person, In Bhilwara, Barmer, Bharatpur, Jalore, Jhalawar Nagaur, Rajsamand and Sawai Madhopur districts wheat distribution was more than 3 quintals. The wheat to such category people is provided free of cost from the Government of India and the State Government is to bear transportation cost for carrying it from Food Corporation of India Jaipur to respective districts.

8.2.7 Nutritional Support to Children and Pregnant Women

Nutritional support was provided in the drought affected districts through Integrated Child Development Programme and Mid-day Meal Programme. Under ICDS programme, all the 237 villages were covered and also 20 urban blocks. This is integrated effort of Government of India, World Food Programme, CARE and Government of Rajasthan. Under this programme, children of 0-6 age group, adolescent girls, pregnant women and lactating mothers were provided supplementary nutrition through network of 33000 Anganbari centers and also the activities relating to cumulative development of the children of age group 3-6 years. Nearly 36 lakh children were benefited with the programme and Rs. 105 crores were spent due to drought relief measures whereas the normal expenditure under the programme remains at Rs. 58 crores. This was possible with active support of the all associates of the programme.

Midday meal is carried out in about 73000 primary schools of the state, but in view of the severity of drought, this programme continued during summer vacations and additional expenditure of Rs. 120 crores was incurred on the programme arrangement.

Drought is recurrent feature of Rajasthan and relief measures are carried out in the affected areas. The analysis of the drought 2002 reveals that there are various superficial expenses in the name of drought, which can be curtained to significant level in case drought proofing arrangement are carried out in true spirit. The state is receiving significant funds from the centre for drought proofing measures, but these are not implemented in proper manner. Desert Development Programme and Drought Prone Area Programme are being implemented in the state for more than two decades but the problems of implemented districts have not recouped to least extent. There is usual practice to aggravate the problem when it is faced and no permanent measures are taken to mitigate any of the problem.

The state is severely facing ground water depletion to most serious extent but state government could not take any concrete measure to decrease the problem. Water is most precious commodity for Rajasthan but its wastage by the people who own the resources is maximum. The well off people dug their tube wells further to get water which affect all the surrounding area but people are silent spectator of such drastic acts. The State Government is fearful to bring water legislation in view of feared protests from the people who own most of such resource. The severity of drought is the resultant impact of this problem.

Large scale deforestation has imbalanced the environment and ecology of the state but no efforts are visible to restore the vegetative cover. Even he forest areas are not safe from this problem and most of forests have become degraded in view of their helplessness. In addition, there is no control over the encroachment over the forest lands by powerful people and such lands are regularized in their names in due course of time. This disastrous act has created uncertainty of monsoon and vegetative cover is decreasing continuously. The dryness of the atmosphere is increasing creating serious problem in cultivation practices. The productivity of land is seriously affected in view of various degradation measures. Therefore the severity of drought is resultant impact of human interface which need to be checked through awareness generation and active participation of the people to reverse the phase.

9

Environmental Sustainable Development Plant of Drought Management

The environmental and socio-economic impact assessment of drought has been analyzed with multi-faceted aspects. In is certainly true that drought has become recurrent feature for Rajasthan state in view of its possibility due to failure of monsoon. Total economic of the Government of Rajasthan has been dwindled in view of its occurrence. One severe drought of wide spread nature completely disrupts the environmental status which has already degraded to significant extent. In severe drought conditions, state economy is further deteriorated as relief measures require voluminous funds to help the affected population and the state is further gripped into loan to be repaid in future. It is true that the Government of India provides assistance to manage relief operations, but this assistance has 70 percent loan to be repaid along with interest. Rajasthan is agrarian economy and most of the cultivation practices are based on rain-fed practices. Therefore, every drought puts a setback to the state and people of the state.

Drought is caused due to failure of monsoon, which is common practice in the state and there are very few years in the past when total state remained free from the drought. Drought has been treated as natural factor and monsoon plays important role in occurrence of drought, as it has leading role from beginning to end. Failure of any of the spell is dangerous to the economy of the state as crop failure is the resultant factor of such happening. Rajasthan state is largest in the country covering 10.40 percent of country geographical area and the state has varied geo-physical conditions. Each agro-climatic zone has its special characteristics different from others. The behaviour of monsoon remains varied over the state and one or another part of the state remain affected with erratic attitude of monsoon. The monsoon has very erratic nature from beginning to retreat and people of the state remained feared from its drastic impact.

Drought has multi-faceted effects over the state economy where most critical issue is cultivation system, where most of the area is solely dependent on rain-fed practices. Even the cultivated area can increase by 60 percent in case of widespread rainfall and in case of regular and widespread rains, the crop production is bumper, but the production is affected drastically in case of monsoon failure. In such erratic conditions, the cultivators are to survive. With all the extremities of nature, rate of population growth is highest and livestock population is more than human population. The human interface has deteriorated the environment the state to severe extent and socio-economic conditions of the people are miserable.

Natural factors are responsible for drought conditions in the state but the human interface has created most of the natural adversities. Uncontrolled felling of trees for human requirements is most critical factor which has denuded the earth surface. Clearance of vegetation for cultivation purposes has further lost the vegetative cover causing severe damage to environment and ecology of the state. Soil erosion through water and wind has lost the productivity of land and removal of trees has further helped in such drastic acts. Various efforts made for reforestation could not yield success due to human acts and humidity of the land has been seriously affected. Extremity of temperatures have extended dryness in total area of the state.

Drinking water supply to rural areas has been provided with ground water resources, which are very meagre. Technological developments helped in exploitation of ground water as part of green revolution has depleted ground water resources to critical extent. Total state has been reached to most dangerous level of ground water depletion where further draft of ground water may extinct these resources completely and such situation is started in some parts of the state. The traditional system of rain water harvesting has been completely disrupted which has further damaged the situation. Salinity and alkalinity has degraded cultivated lands and most of the fallow land has become unfit for cultivation practices. For feeding the growing population, use of chemical fertilizers and chemical pesticides has helped in increase of agriculture production and productivity, but natural fertility of the land has been drastically diminished. Half of human diseases are the impact of use of chemical fertilizers and pesticides and various species of birds and animal are in the process of extinct.

The erratic and uncertain nature of monsoon is mixed impact of all the factors of human interface. The position has deteriorated to such extent that even small deviation in rainfall effect over the cultivation practices and being agrarian economy, total state gets setback in case of failure of monsoon. The state is supposed to provide relief to the people due to occurrence of drought, but how long this process will continue. The extent of land degradation and desertification has reached to most critical stage where the productivity of land is regularly decreasing. The imbalance between increase in population and decrease in agriculture production is bound to create most disastrous condition of this state and there is feared situation of becoming state like Sahara or Sahel deserts.

Droughts are become famine in view of serious environmental degradation and socio-economic conditions of the state are deteriorating continuously. The efforts made to revive the situation could not yield any successful results to decrease the intensity and magnitude of drought. Thus drought is the most severe disaster to the state, where retrieval measures are not succeeded to significant extent. The state government claims to grow more than one crore trees each year and claims plant survival rate of more than 70 percent but even after thirty years of plantation, no visible impact over the state is there. Similarly the impact of desert development, drought prone area progremme, reforestation of forest areas, watershed development and treatment of waste and barren lands could not yielded much.

Thus development and destruction cannot run simultaneously as the people and the state government needs to choose one path by banning another mode completely and effectively. If the state government wants to carry out drought proofing measures, the destructive acts need to be stopped completely and that should reflect in its functional attitude. The deterioration of environment and ecology can be possible with curbing human acts, which must reflect into action. The forest areas are fully protected with legislative measures and if the state government is unable to stop further squeezing of forest land by stopping unauthorized acts of capturing of forest land, other lands owned by the government is unable to be checked from illicit occupation of such lands. The position of cultivated lands is deteriorating on continuous basis and without treatment of such lands cultivation practices would not remain remunerative to cultivators.

The severity of droughts would continue to remain disastrous as all the conditions are prone to recurrence with drastic impacts and magnitude. In this situation there is necessity to think over the future consequences of droughts as every drought is further disastrous to the state and there is necessity to take concrete measures to decrease recurrences of droughts for socio-economic development of the state. The people of the state are leading most miserable life and cultivation practices are becoming lesser remunerative on regular basis. Therefore, there is necessity to have complete stocktaking of the present status of the state from environmental and socio-economic point view in the recurrence of drought conditions. The years of 1997-98 to 2002-03 remained most severe drought years while 2005-06 and 2006-07 had also drought conditions with moderate severity.

The environment degradation has become regular feature, as human interface is increasing continuing. There are various spheres where the level of environmental and ecological degradation has reached to such level where all the factors may become dysfunctional in recent future. The land surface has been denuded to vital extent and deforestation is uncontrolled feature. Encroachments on forest land has squeezed the area and even protected forest regions are denuded. Open grazing by livestock has made the earth surface without vegetative cover. Level of ground water has reached to most critical level where further draft of water may extinct these resources. Natural fertility of cultivated lands has been lost due to excessive use of chemical fertilizers and chemical pesticides and for crop yield at least 30 percent more water is required. All these factors have deteriorated the natural system of environment and ecology, resulted in to erratic and uncertain rainfall over the state regions.

The severity of drought has been increased the disastrous impact where famine conditions prevail over the affected areas. With large scale deforestation, various species of wild life has been extinct while other are at the verge of disappearance as their shelter, food and water has been snatched by the people for meeting their own requirements. Deforestation is the only single factor disturbed the natural environment and uncertainty of rainfall is the resultant impact of this very factor to vital extent. In view of the present level of human interface, the drought is further creating conditions of devastation. In view of these factors, the following features are feared to intensify the problem, if remedial measures are not started with immediate effect.

(a) Uncertainty of rainfall may cause recurrent droughts creating serious conditions for survival of people in the state. In addition to drought, floods may also cause damaging acts as the soil structure of the state is unable to bear heavy downpour. Replenishment of one widespread drought takes at least three to five years and recurrent droughts may cause serious concern to the environment and ecology of the state.

(b) Large scale deforestation has denuded the earth surface in most part of the state, while remaining vegetated land may also cause similar conditions in view of uncontrolled human acts. Forests help in maintenance of humidity and moisture in the earth surface and atmosphere and denudation may cause serious soil erosion through wind and water.

(c) Ground water depletion has reached to most critical level and almost total state has been converted into dark zone and most critical level where ground water resources have started extinct process in some areas, while other areas are in the process.

(d) Clearance of vegetative cover of earth surface for meeting the food requirements of the people. The cultivation practices initiated in deforested lands give high yield due to presence of moisture but it is lost in few years for remaining unused for one season or remaining fallow for few years. Presently half of the cultivable land is compelled to left fallow which has lost its fertility in view of loss of moisture and there is serious problem of salinity and alkalinity in larger area of some districts. Such conditions cause due to draft of saline water from ground resources.

(e) Natural fertility of soil is diminishing on continuous basis due to excessive use of chemical fertilizers and chemical pesticides and excessive doses of such chemicals have become inevitable to maintain productivity level. This will create serious problem of decrease in production.

(f) With occurrence of monsoon, the bare land is covered with grass which is grazed by livestock and total land surface becomes barren again. Soil crest with livestock unchecked movement is flown away by windstorms and the fertile soil cover is lost. Grasses grown on natural conditions can be helpful to provide nutritional support to livestock if collected. No efforts have succeeded to develop pastures on permanent basis.

(g) Growth rate of human population is much above the national level and it is difficult for the state to meet food requirement of population in coming years if present level of degradation continues.

(h) Livestock is dependable source of livelihood for many people but livestock population is more than human population. Though livestock is helpful for creating dependable survival base to the people but the environmental degradation is most vital factor due to uncontrolled grazing and movement of livestock.

These are few factors causing serious concern for deterioration of the environmental conditions in the state and there is immediate necessity to consider whether such devastating factors can be checked or human interference may be allowed to play its role to deteriorate it further. Droughts have become recurrent feature in the state and drought measures being taken up in the state must be result oriented and fully determined to reduce the severity of the drought in coming years. The behaviour of monsoon can be changed if human interface is stopped to maximum possible extent. The drought mitigation measures to be carried out after assessment of local conditions and all the factors responsible for drought and unfavourable monsoon need to be corrected in phased manner. This process is tough and may take several years but once the revitalization process initiated should continue forever as mandatory condition.

9.1 SUGGESTIONS FOR SUSTAINABLE DEVELOPMENT OF DROUGHT MANAGEMENT

The best way for revival of environment and ecology of the state, main role is to be performed by the people of the state and awareness generation is the necessity to converse them from the disadvantages of drought conditions. The causes of drought need to be narrated in detail along with the message that the natural factors responsible for drought are the natural factors, which are creation of human acts. Short term measures and long term initiatives need to be spelt out for each agro-climatic zone with clear elaboration of factors responsible for creation of imbalanced environmental conditions. There is long list of corrective measures which the people and the government need to perform for the benefit of the mankind and state regions. The suggestions given for rectification of the environment and ecology as well as socio-economic development will not cost any additional burden on the state but can be carried out as part of ongoing central and state schemes. These suggestions are easy to be adopted by the state with peoples participation but there are some conditions, which the people of the state and the government need to perform without any interference. The pre-conditions for drought mitigation and environment development are given hereunder:

(a) No live tree will be cut in the state by any one within the habitations or by outsiders. There are some regions in the state where trees are protected and no one is allowed to damage live trees for any use.

(b) Uncontrolled grazing need to be stopped and every person possessing livestock will form habit for stall feeding.

(c) Ground water will be used for meeting drinking water and irrigation requirements only up to the level it is annually charged and stress will be over water harvesting to maximum possible extent. This process is necessary for serious dry zone regions.

(d) Use of chemical fertilizers and chemical pesticides will be stopped at the earliest and every cultivator will use bio-fertilizers and pesticides for meeting total requirements.

The pre-conditions mentioned above need to be strictly followed and success of other measures will depend on the pre-conditions. There is no fear for any mishappening as every aspect has been addressed in befitted manner and able to mitigate all the serious problems being faced by the people and the government. The total suggestive measures can be completed within a period of five years. The droughts would continue as natural process but their intensity and occurrence would be reduced to significant extent. The state will be able to regenerate its vegetative cover and people of the state will prosper. Awareness generation is the important feature of this suggestive package and people of the state need to be associated in implementation of these programme with empowerment in total decision making process. In addition there will be enormous employment opportunities for the people who are presently struggling for their survival in view of recurrent droughts. There is no reason to fear from the proposed pre-conditions as every necessity of the people of the state has been suitable addressed in package of suggestions. The suggestions are combination of drought proofing measures and other issues necessary for revival of environment and ecology of the state.

9.2 LAND USE MANAGEMENT

The land use pattern in the present system is most irrational in view of serious degradation and various other problems. The land resources are limited and pressure of growing human and livestock population has deteriorated the land to most severe extent. The unutilized land has become unproductive in view of human interface and no efforts for revival of its use could be possible irrespective of various schemes. The land use pattern of 2002-03 reveals following set up:

(a) 26.51 lakh hectare or 7.74 percent area: Forest land under reserved, protected and unclassified categories.

(b) 17.64 lakh hectare or 5.15 percent area : Land not available for cultivation for being used in habitations, buildings, railways, roads etc.

(c) 25.14 lakh hectare or 7.34 percent area: Barren and uncultivable land.

(d) 17.15 lakh hectare or 5.00 percent area : Permanent pastures and tree groves.

(e) 48.66 lakh hectare or 14.20 percent area : Culturable wasteland.

(f) 32.59 lakh hectare or 9.51 percent area: Fallow land other than current fallow.

(g) 66.88 lakh hectare or 19.52 percent area: Current fallow land.

(h) 108.07 lakh hectare or 31.54 percent area ; Cultivated land.

The present status of forest land is most pitiful as forest department is unable to protect the land from encroachments. The problem of encroachment is most common as people manage to capture such lands for meeting their cultivation requirements and later these lands become unproductive land. The forest land of 7.74 percent of total geographical area is not sufficient for environmental and ecological point of view. The United Nations Convention for Combating Desertification (UNCCD) has suggested putting up 33 percent of the geographical area under plantation activities. This is also necessary from environmental point of view. In the present set up of land use pattern, 7.74 percent forest land and 5.00 percent permanent pastures and tree groves is available for vegetation activities.

While making 33 percent land under vegetation, it is also necessary to address peoples requirements of fuel wood, fodder and timber requirements as every aspect need to be addressed in befitted manner. The present problem of uncontrolled felling trees is the resultant impact of peoples requirements, which need suitable addressal as people of rural area need wood for their various requirements and unless these requirements are fully met, felling of trees will continue. Therefore, while making the strategy for vegetative cover over the earth surface, it is necessary to address all human requirements for survival of vegetation cover on permanent basis. For this purpose, the tree plantation need to cover timber, fuel wood and fodder requirements and this is the only way to protect trees over the earth surface.

Presently the area under different categories available for plantation activities is 34.28 percent of the total geographical area and 1.28 percent area need to be left for various land requirements like urban areas, industrialization and other developments in addition to present 5.15 percent area. Thus planning for remaining 33 percent of land for plantation activities need to be done in most rationale manner to meet all human requirements. The present forest area is mostly degraded, which need to be increased to at least 11 percent level where plantation of timber can be successfully done. Roughly 51.40 lakh hectare area need to be put under timber plantation with local adjustments. Timber plantation is possible at the rate of 1250 to 2500 tree plants per hectare, which means that 64250 to 128500 lakh trees of timber can be raised in proposed forest areas.

The forest area need to be protected as per present system as this area will help in providing shelter to wild life, which is also essential feature of the environment and ecology of the area. Gap between tree plants is maintained of 2 to 4 metre per plant for development of trees and gap area can be used for plantation of grass as that will help in meeting the fodder requirement and such grass need to be sold to people at very nominal rate to be born by the livestock raisers. This amount need to be preserved by forest department for maintenance of forest land. Wood of dead trees can be sold as per practice of forest department, but in the transitional phase, there will be vacuum for meeting the timber requirement which can be met from outside the state as preservation of such trees is most essential.

Another 11 percent area for vegetative cover need to be covered under horticulture trees with same amount of trees. Such tree plants need to be protected by people of village community Even minimum 64250 lakh trees of fruits are raised in barren and wasteland and handed over to the village people who are willing to shoulder this responsibility. Twenty five fruit trees to each family are provided over the waste and barren land after treatment of degradation problems, 2570 lakh families can be benefited with supplemental income generation, which would provide supplemental income to cross the poverty line on sustainable basis. If minimum income from one tree of fruits is Rs. 1000 per tree, all the associated people would get supplemental income of minimum Rs. 25000 per year and this income will be additional over the existing income.

While making such allotment to the people, the rights are given for the tree product and not the land. Any person so benefited need to be made ineligible to sell trees to any other person. These benefits are in lieu of the care and watering by the assigned families, which they will continue in further duration in their own interest. Thus each rural family will be provided additional employment opportunities with assured benefit. However, the assigned families need to be made free to select fruit tree variety keeping in view adaptability in environment conditions and marketability of the produce. The gap area can be used for raising grass for meeting their livestock requirements and this will be added advantage to each willing family.

The remaining 11 percent area can be used for trees of fuel wood and fodder requirements. Usually branches of such trees are cut once in year to allow tree growth in further duration and such wood is helpful in meeting fodder and fuel wood requirements. Similarly 25 trees per family can be allotted to each family to meet their requirements, without harming tree trunk. Thus total rural population will be benefited with this preposition and their wood and fodder requirements would be fully met There are presently less than one crore rural families and each family can get even more trees for protection and benefit. There will be area specific problem in demarcation of plantation area, but such issues need to be addressed at local level.

Plantation of all categories of trees need to be done in the wind direction to protect cultivated land and moisture retention. Large scale plantation is helpful in decreasing dryness of the atmosphere to vital extent and if all the plantation activities are carried out in the assigned areas, the occurrence of rainfall will be less erratic and widespread rains would start in few years. This process will help in soil erosion, wind storms and movement of sand dunes, which is serious problem of desert regions. Process of evaporation would decrease considerably and extremity of temperatures would be controlled to significant extent.

Present cultivated land, old fallow and current fallow lands combined together constitute 60.57 percent of state geographical area. Most of the land is marginal and degraded with decreasing productivity. Treatment of total cultivable land is the necessity to utilize for productive use for meeting human and livestock food and fodder requirements. Various processing industries are functioning in the state in the hope to get sizeable task but due to recurrent drought conditions, crop damage disrupts total process. The cultivation practices in all the agro-climatic zones of the state is possible with the help of moisture retention through plantation activities. There are various causes of land degradation and decrease in the productivity yield. Total cultivable land need soil treatment by meeting all the existing deficiencies for the sustainable productive use of these lands. The plantation area of 33 percent would address most of the problems of cultivable land if proper treatment is made. Land degradation and desertification are the serious problems of the state, which need careful and sincere efforts and once the treatment of land is carried out, people engaged in cultivation practices would also understand treatment process. The treatment tasks need to be attended by the state government.

There are various issues of land degradation and treatment practices need to be attended at local level as local problems are based on geo-physical conditions. Cultivated and plantation lands of Ganganagar and Hanumangarh are seriously affected with the problem of water logging and this problem is common in canal irrigated lands where cultivators never bother to save water. This problem also exist to some extent in Kota, Banswara and Dungarpur districts. Salinity problem exists most in the areas where ground water is saline or draft from very low depth. Such areas need to be cultivated with rain water harvesting or through moisture retention as per feasibility conditions. In some regions, alkalinity problem is most critical and large area has become uncultivable due to this problem. Such lands need suitable treatment with gypsum for constant application of three to four years to regain the soil fertility.

Cultivators of state need to realize the acute shortage of water from all available resources and therefore, the cultivation practices need to be carried out with minimum water requirements. There are various water saving devices, which can provide similar yield with minimum water. In unirrigated areas, short duration crops need to be encouraged to get remunerative return from cultivation practices. In addition there are various drought resistant varieties of seeds have been introduced to grow crops in erratic rainfall conditions. Occurrence of rainfall is natural phenomena and would become favourable after some years but meanwhile the cultivators need to manage their cultivation practices in feasible manner where chances of crop damage can be minimized to possible extent.

Up to present level the state is self sufficient in food grain production and when total cultivable land is possible to put under agricultural activities, the cultivators will be required to shift to commercial cultivation practices for getting remunerative return. Cultivation practices in about 40 percent of the area under food grains would suffice the requirements of the people and rest 20 percent area need to be put under medicinal and commercial cultivation with crop varieties able to be grown in dryland conditions with minimum water requirements.

Moisture cultivation practices carried out in Jaisalmer district from centuries are most befitted for Rajasthan where density of population is low. There is need to spread the area in Jaisalmer, Barmer and Bikaner districts where rain water run-offs are possible and cultivable area is significant. Such practices are also possible in other regions where run-off water can be harvested or directly diverted to cultivated sloppy fields. After large scale plantation, many areas of the state would be covered under moisture cultivation practices in view of favourable atmospheric conditions. Crop harvesting practices can be suitably designed on local planning system keeping in all the factors in view.

The agro-climatic zones of the state are drawn on the basis of geo-physical and atmospheric conditions and separate strategy for each climatic zone need to be drawn in most befitted manner. This strategy will help in decreasing the severity of drought conditions in each region and crop damage would be minimized to great extent. The replenishment period of drought would also be decreased as the vegetative and moisture loss will be minimized considerably. It would also be possible to limit the drought to becoming famine in most parts of the state with application of the given framework.

9.3 WATER MANAGEMENT

Water is the most critical factor which can be clarified with the fact that the state has 10.41 percent of the country geographical area, 5.50 percent of country population and only one percent of country water resources. In this reality, the strategy of water management needs to be drawn for sustainable development of the state and decreasing the impact of drought to maximum possible extent. Water management is most challenging task but this can be practiced with most careful planning, which need to be framed in most critical conditions as total state area is seriously facing the ground water problem and almost all the regions of the state have been converted into dark zone and serious critical zone due to excess draft of water from ground than its recharging capacity. This is a matter of great concern for the state as a whole as further draft of water for meeting irrigation requirement is dangerous for extinction of these precious resources.

The state government is helpless to check the process of drawl of ground water as irrigation is the necessity for agriculture production and ground water resources have depleted to drastic level. Irrigation system is seriously affected with failure of monsoon as the variance of irrigation ranges between minimum 43.72 lakh hectares to maximum 56.12 lakh hectares. In view of favourable rainfall year of 1999-2000 the irrigation was possible in 56.12 lakh hectare area, when the net cultivated area was 155.09 lakh hectares. In the worst rainfall year of 2002-03 the net cultivated area decreased to 108.07 lakh hectares and similarly net irrigated area was also decreased to 43.72 lakh hectares. This variation is significant for the state and cultivator community and both suffer to most serious extent.

In the year 1999-2000 when net area irrigated was 56.12 lakh hectares, 28.84 percent area was irrigated from canal water, 1.40 percent by tank irrigation, 68.90 percent area was irrigated through wells and tube wells and remaining 0.84 percent by other sources like anicuts etc. In the year of 2002-03, the most severe famine year, irrigation was possible in 43.72 lakh hectare area where area irrigated through canal system remained 12.96 percent, 0.18 percent through tanks, 77.24 percent through wells and tube wells and remaining 0.62 percent through anicuts. This is most critical phase of the cultivation practices and cultivators try their level best to make tough fight with monsoon and nature. All the three sources of irrigation like canal, tank and anicuts are part of surface irrigation system but the ground water irrigation has reached to a critical level varying from 68.90 percent to 77.24 percent, which is evident of pressure on ground water resources in the state.

The problem of charge of irrigation water in different soil strata of the state is serious as 20 percent of rain water coming over the earth surface is charged to ground water resources. Thus rain water harvesting over the earth surface to maximum possible extent is necessary and part of the strategy of water management. This water is necessary for meeting the irrigation and plantation requirements if water is collected in open areas. Similarly rain water in habitation lands can meet drinking water requirement to significant extent. The total water management strategy is based on this system to decrease the impact of drought to least possible extent. This is the only measure for sustainable development of the state with assured drought proofing measure and also for vegetative cover over the earth surface of the state.

The proposed strategy will be able to harvest rain water up to 40 percent from the earth surface and remaining water would continue to charge ground water resources. There is limitation of rain water harvesting process as it is possible to harvest rain water from very few strategic points, but would suffice to meet the total strategy areas to maximum possible extent. In Rajasthan, rain water harvesting is possible from hard structures like mountain and hills, pavement and pediments where rain water runs to lower regions and run offs are created. Such efforts are also possible in sand dunes regions of more than 9 metre height and this process is most feasible in desert regions as rain water runs to lower regions and charge inside the earth surface in slow process. Most of the khadins used for moisture cultivation practices are managed through this process in the dunal regions of significant height and water harvesting is possible to manage both plantation and cultivation practices.

Rain water harvesting of forty percent level has also been planned in view the fact that present flow systems creating significant run-offs and water harvesting on large scale need not be disturbed as this process is continuing for many years and disruption in such systems is not proper. There are many khadins functional for centuries and such practices are also part of water harvesting practices. Thus rainwater is possible from untapped sites for meeting the additional requirement of plantation and cultivation sectors as per harvesting level. If rainwater harvesting is carried out in the desired manner in habitations and open fields in permanent structures for drinking water and plantation activities, most critical part of the strategy is able to be fulfilled.

Management of harvested rain water is to be done in most judicious manner to get maximum return. For plantation activities, the total process of plantation is to be carried out in phased manner as total proposed plantation in 33 percent of area is not feasible in one stroke. Therefore phasing of the plantation activities is necessary. There is most critical and ambitious target to carry out plantation activities in 154.20 lakh hectare area where minimum plantation work would be to raise 38.55 crore plants in total state region. In this process, first task is to select feasible sites in untapped areas, which can only be possible with the help of local people at village level as village is to be unit of implementation. There are more than 41 thousand villages in the state and in the first phase of the strategy only 10000 villages will be taken up for selection of most feasible sites of rain water harvesting.

The rainfall trend of the state reveal that there is at least one rainfall with heavy showers even in most deficient years, and rain water harvesting is possible to minimum The rain water harvesting structure constructed at feasible sites can be filled in as per rain water charged through run-off and the rain water need to be cemented to retain water without any seepage or evaporation losses. Planning for collected water is to be done by the peoples association keeping in view the requirement for plantation activities for whole of the year and specially during the summer, when water requirement is maximum. Remaining water can be used for meeting the irrigation requirements where sharing of water among users need to be done in most reasonable manner where the village association have to play active role. Plantation water requirement remain for three years when the tree plants become self manageable to their water requirements. Even after the plantation activity is fully geared up, tree plants would require more water in horticulture trees, which need to be addressed in further duration.

Thus each village community where water harvesting structure is constructed at feasible site and use of water is started in most rationale manner to grow more plants, the environmental and ecological conditions are bound to change where the intensity and magnitude of drought would start decreasing slowly but steady manner. After every three years, water would be available for new plantation sites and total area of 33 percent would be covered with vegetative cover in phased manner in total duration of nine years. The people of rain water harvested sites would start witnessing the changes in atmosphere and ecological conditions of the total geographical area in terms of extremity of temperatures, availability of humidity and moisture in the atmosphere. This change is most favourable for the vegetation and cultivation practices where the frequency of drought start decreasing.

In the regions where rain water harvesting is practiced, people of the habitations need to be motivated through awareness generation to use minimum ground water for irrigation requirements to allow repletion to the extent when the fear of ground water depletion is checked and made stable. This is the situation, which is part of the strategy of the state regions. It is envisaged that within a period of ten years, the changes in occurrence of rain and its erratic nature would be checked. Water saving device use of all plantation and cultivation needs is mandatory and surplus water need to be conserved or retained for future requirements. However, such situation would occur in good monsoon years.

9.4 DROUGHT PROOFING MEASURES

Funds received for various drought measures to the state need to be used for construction of rain water harvesting structures. Every year, sizeable funds are received under Desert Development Programme, Drought Prone Area Programme, Wasteland Development, Reforestation and plantation activities under these schemes and other various schemes need to be used only for construction of rain water harvesting structure. The State Government need to create awareness generation, empowerment of village people in total decision making process relating to selection of sites for rain water harvesting structures and plantation activities. The people are well versed with the sites where run off water can be carried to such structures to the maximum possible extent. The government need to concentrate over the construction of rain water harvesting structures with people participation and plantation activities need to be carried out by needy persons who willingly express their desire to maintain the assigned number of plants. The government departments can help in providing tree plants of desired species, treatment of land where plantation activities are to be carried out. Total process is to be carried out in most coordinated manner where government departments should work in accordance with the decisions taken in village association meeting and name of the people intending to participate in the programme and all category land to be used under plantation activities need to be allotted to village development association with clear understanding that the land used for plantation activities would continue to remain the government or panchayat property and people will be allowed to share the benefits, without harming trees and assigned area.

The activities for drought proofing need to be changed into the strategy made and if there are some problems from the Government of India side, total plan need to be sent to them for their concurrence for total plan of drought proofing. The strategy framed for Rajasthan state cannot be implemented into piecemeal or as per the guidelines of various schemes. The state government need to approve total process in terms of cost and application and made clear to the Central Government that construction of rain water harvesting structures is their first priority and remaining activities would be carried out by the village people as per state guidelines where sufficient flexibility is to be given to local people in terms of their geo-physical conditions but they would strict their activities to these areas. People of the state want water from any source and water is made available they would manage other activities as per state guidelines.

Space technology can be also helpful in selection of suitable rain water harvesting structures and such units would prove result oriented. The sites selected through space technology in respect of each village need to be reviewed in village development association meeting and most reasonable sites would be selected with the technology support and peoples perception and flexibility in selection of sites for plantation and water harvesting structures is necessary and there cannot be any possible formula designed at the state level. The success of various activities implemented in the past were not possible in view of the fact that the design was faulty and there was no association of the people in decision making process which is the main theme of the success. The activities of sand dune stabilization could not succeed as people started harming trees and bushes planted by the government functionaries.

Success of drought proofing measures in only possible when people are involved in their own interest. The oran culture of Rajasthan flourished only as people treated these sites as their own and once ownership is developed among them, people would not allow anyone from village or outside to harm their trees. There is no harm if people think on those lines on which orans were established. Various issues of environment and ecology are part of the ethnic values and if technical support to their own cause is managed, every person would be part of these activities in villages. Selection of the people can be done after close assessment of people's desire and every person opted to carry out the drought activities need to be finalized after thorough assessment of its strategy.

The action plan for drought proofing need to be redesigned as there are various areas, which need not be covered under state implementation. All the activities to be taken up under drought proofing under any of the programme should focus on rain water harvesting to maximum possible extent. Plantation activities are to be left to be done through village people under the guidance of the experts of respective discipline of state government Soil conservation measures need to be done in arable and non-arable land for stabilization of land and safety of the soil displacement from present locations. The cultivated lands also suffer from this problem to great extent. The government must be vigilant for obtaining the benefit of the central sector, centrally sponsored and special assistance programmes for this task. Development of nurseries is necessary for assured availability of plants and people need to empower for selection of plant species from government or private nurseries from any reputed institution where more than 90 percent of plants remain live.

Activities to be managed by the people of village community need to be initiated by the state government as demonstration project where the wishes of the people need to be honoured as they need quality plants with maximum survival rate and better yield. In the schemes to be implemented through peoples participation, resource generation must be the responsibility of the people in terms of cash, material or wages. Presently this participation is envisaged up to 30 percent of the total cost. This is essential to develop ownership in the people by seeking their participation and people would take their own decision in the beginning or after viewing the advantages of various drought proofing activities, The procedural formalities for approval and other issues related to nature of participation and contribution of the village community who would provide their share as per their capabilities. The behaviour of government functionaries with the people must be associative and responsive.

The ongoing state programmes have provision for drought proofing measures and the expenditure need to be made on the activities, which need to be carried out by the people at their own. In cases where people are willing to pay in their own interest, such activities need to be carried out with peoples participation. Such empowerment would help in sustainable manner. The ultimate objective of available funds under drought proofing measures to be spent over creation of permanent structures keeping in view the availability of water to these units. The peoples participation would also be helpful in solution of unrest and resentment in the people if there are some affected people using same water. Every proposal submitted before the village development association will be finalized from every angle and there would be no local problem as all such issues would be solved at the village development association at grass root level. If there are some local arrangements to address local problems which sometimes crop up as resentment of some beneficiary groups, who are already the benefits of similar nature.

9.5 SOIL MANAGEMENT

In the present, natural fertility of soil of arable and non-arable land has been degraded to most critical level, causing serious concern to production and productivity. About 70 percent of the land of arable and non-arable category is degraded and marginal and regaining the fertility of the soil is most critical but non impossible as various research organizations of the state and non-government possess capability to treat all sort of deficiencies of the land. On assured availability of water for plantation and irrigation of at least existing level, treatment of soil is most important task to be performed. The mind set of the state government functionaries rests with the approach that working on private land is the responsibility of cultivators. Similarly treatment of wasteland, forest land, barren land and pasture land is not feasible in view of serious water shortage and even protection of forest land is not possible.

Government measures to treat land even for plantation activities is never considered as most important measure for drought mitigation. Similarly, the success claimed by research institution for treatment of soil deficiencies never attempted over the cultivated fields though the government is well aware of decreasing production and productivity. The main hindrance in development of Rajasthan is the availability of water and ground water presently used for irrigation and drinking water needs would not remain available in coming years. When the water will be available to meet the human requirements, soil treatment for rain crops and post rain crops would be possible with harvested water and soil moisture. Even the rain failure could be possible to mitigate to some extent with harvested water in the monsoon duration.

Soil management would require treatment of the deficiency of the soil like salinity and alkalinity as well as water logging. For all category of deficient soils, water saving devices need to be used as soil of the state is unable to retain excess water. Water logging problem is most recent in canal irrigated lands and causing serious problem to soil fertility. The soils of the state can be most fertile with proper treatment and able to provide moderate yield if water is available to requisite extent. The plantation activity would help in retention of moisture and such practices are already being practiced in desert regions of the state. Plantation over the earth surface would change the adversities of atmosphere to great extent and impact of drought would be mitigated to vital extent.

Another problem being faced in the state is diminishing natural fertility of soil due to excessive use of chemical fertilizers and chemical pesticides. These chemical require about 30 percent excess water to save crops from impact of chemical fertilizers. This much of water is most critical to water deficient state like Rajasthan. For maintaining the productivity level higher doses of chemical fertilizers are given which require more water and natural fertility of the soil. For this purpose, bio-fertilizers are the best alternative as this process would save more than 30 percent of water for irrigation and would be helpful in depletion of ground water to great extent. Preparation of bio-fertilizers is possible at the village level with local resources which is used as fuel for food preparation. The produce with bio-fertilizers and bio-pesticides would maintain the natural fertility of the soil in sustainable manner and there is possibility of 10 percent addition yield. The cultivators will get about 15-20 percent additional price of agriculture produce.

Use of chemical fertilizers and chemical pesticides has been banned in developed countries in view of environment degradation and adverse impact over the human health. About 50 percent of diseases in human beings are due to chemical fertilizers and chemical pesticides. The loss to environment and ecology is most severe as various species of wild animals and birds have extinct and other are in the process of extinction. The soil fertility of most ports of the state has lost its natural fertility and there is immediate necessity to save it from further degradation. The plantation over land surface would help in saving the wild life as those will also get shelter and food and cultivated lands would regain natural fertility to provide food and other produce to population. With this strategy, the impact of drought would be minimized to vital extent, as soil will regain its natural fertility.

9.6 Human Resource Management

The present human resource management of Rajasthan in rural sector is severely affected with the drought disaster. As per Census of 2001, out of total rural population of the state of 432.68 lakhs, 139.58 lakh persons are main workers, 59.20 lakhs are marginal workers and 233.90 lakh persons are non-workers, which constitute 32.26, 13.68 and 54.06 percent of rural population respectively. Out of main workers, 94.07 lakhs are cultivators and 9.65 lakhs are agricultural labours. Under marginal workers 35.14 lakhs are cultivators and 14.71 lakhs are agricultural labour. Thus main and marginal cultivators and agricultural labours combined together constitute 35.49 percent of total rural population. Total main and marginal workers in rural Rajasthan are 45.94 percent of the total population, which are 77.25 percent of total rural working force Working level of persons engaged in agriculture as cultivators and agricultural labour is well known as about 73 percent cultivable land is possible to be cultivated under monsoon crops and chances of failure of rainfall remain quite frequent.

With this background of the human resources, the management of human resources is most essential. The working population of rural Rajasthan is 53 percent and planning for total population is necessary for saving from clutches of drought. Livestock rearing is one of most dependent occupation of rural people as this helps to face the adversities of monsoon to vital extent. Monsoon failure is regular feature of crop damage and there is necessity to associate the rural population with sustainable employment opportunities as part of human resource management. This is most essential to save people from drought disaster and plantation activity can be most suitable for the people of the state specially to face the drought situations. With rain water harvesting and plantation activities, total working population of rural sector will be able to get supplementary employment better than livestock rearing and with both the activities the people of rural Rajasthan will be able to face the vagaries of drought effectively.

Water harvesting and plantation activities are able to change the environmental and atmospheric conditions of the state as total state rural population will be able to get sufficient employment opportunities and there will be no poverty in the state. With this strategy every family of rural area will be able to get 25 trees of timber, 25 trees of horticulture and 25 trees of fuel wood and fodder to share benefit of trees. This proposition will help in getting assured income of more than Rs. 20,000 per year which is poverty line for rural people. Income from cultivation and livestock rearing would be additional to these efforts. In addition each family will be able to grow grasses in gap plantation area.

Total strategy for drought mitigation and prosperity of the people of the state is not imaginary as every item of action plan is feasible and there is no doubt for the success. This strategy is based on rain water harvesting method which is possible in Jaisalmer district, where normal annual rainfall is 18.55 mm, lowest in the state. This is surprising fact that people of this district are carrying on moisture based cultivation practices and there are various potential areas of rain water harvesting. accordingly, if the proposed is possible in Jaisalmer district, it is possible in total state as every part of the state is getting normal rainfall up to 95 cms with lowest of 18.55 cms in Jaisalmer. Rajasthan state possesses about 34 percent land under forests, barren land, waste land and permanent pastures and tree groves. Therefore no activity of plantation and rain water harvesting is proposed on the cultivated land. The cultivated land of 61 percent will have more irrigated and moisture area than the present and drought severity on agriculture would decrease to significant extent. The people of the state would prosper with plantation activities and crop damage would be minimized to moderate level. The drought relief works would not be required to be carried out even in severe drought conditions.

So far as availability of funds for drought proofing activities are concerned, the main task of the activities would be construction of rain water harvesting structures. There are more than 40000 villages in the state and average size of such structure would require about one crore for construction of rain water harvesting structure, soil management, plantation and other related activities. Thus out of total requirements of Rs. 40,000 crores, funds to the tune of 10,000 crores can be dovetailed from existing central and state plan funds as these funds are meant for such activities. There is necessity of the preparedness of the government to understand the process of rural development where the impact of drought can be minimized to great possible extent. Remaining funds can be managed under next five year plan as the results of preliminary process would automatically firm up the state government mindset and the benefit to the people as well as the environment development would encourage the state and people to carry forward these task with great zeal.

The changes in the environment and atmospheric factors would help to the people of the area in their socio-economic development and they would gain strength the face the adversities of drought with courage and vigour. The state require water management for its development process and would realize that the Rajasthan state known for desert, drought and damaging environment would change into most vulnerable land. People of the state would not survive in pitiful conditions.

BIBLIOGRAPHY

Archarya, SS (1989), Socio-Economic Impact of 1987 Drought in Rajasthan (ICAR Adhoc Research Project, Final Report) Department of Agricultural Economics, Rajasthan Agricultural University, RCA Campus, Udaipur.

Bhalme, HN and DA Mooley (1979), Large scale droughts/ floods and monsoon circulation, Mon. Wea. Rev. Vol. 108, No. 8, Aug. pp 1197-211.

Bhalme, HN, RS Reddy, DA Mooley and Bh.V. Ramanamurty (1981), Solar activity and Indian weather/ climate, Proc. Ind, Acad, Sci, Bhaskar Rao, NS Vol. 90 No. 3.

Bora AK (2002), Monsoon controls and Brahamputra floods in Assam, in BK Adhikari *et al.* (ed) Brahamputra problems and prospects, Printworth, Guwahati.

Census of India, Paper 1, 2 and 3 (2001), Census Operations Organization, Government of India.

Central Arid Zone Research Institute (2000), Strategy to combat drought and famine in the Indian Arid Zone, Indian Council of Agricultural Research), Jodhpur.

Central Arid Zone Research Institute (2002), Drought Management in Indian Desert, Indian Tourism Publications , Jodhpur.

Centre for Disaster Management, HCM RIPA (2001), Impact of Drought on Livelihood in Rajasthan.

Centre for Disaster Management, HCM RIPA (2006), Drought Management and Administrative Reforms: An All India Perspective.

Centre for Disaster Management, HCM RIPA (2006), Drought Management and Administrative Reforms- An all India Perspective (22-23) May, 2006.

Centre for Disaster Management, HCM RIPA (2006), Workshop on Drought Management, 3rd January, 2006, Drought of Rajasthan- 2002, organized by Centre for Disaster Management, HCM SIPA.

Chanda BN and ON Dhar (1976), Study of incidence of droughts in the Gangetic West Bengal Proc. Ind. Nat. Acad, Vol. 42, Pt. A, Jan.

Choudhary, V (2005), Drought Proofing measures through watershed programme - a case study of Dudu block of Jaipur district, Department Geography, University of Rajasthan, Jaipur (Ph.D. thesis unpublished).

Chouhan, TS (1979), A Geographical Study of the Impact of Drought Hazard and Relief Measures in the Malani Region of Rajasthan, Department of Geography, Jodhpur University, Jodhpur (Ph.D. thesis unpublished).

Chouhan, TS (1981-82), Spatio-temporal analysis of drought and its consequences for the Great Indian Desert: A case study of Jodhpur district, Department of Geography, Jodhpur University, Jodhpur (Post-Doctoral U.G.C. Project).

Chouhan, TS (1982-83), Impact of drought hazard among the rural population in an Arid Environment: A case study of Jaisalmer district, Department of Geography, Jodhpur University, Jodhpur (Research Associate U.G.C. Project).

Chouhan, TS (1987), Agricultural Geography- A Study of Rajasthan State, Academic Publishers, Jaipur.

Chouhan, TS (1992), Desertification in the World and its control, Scientific Publishers (India), Jodhpur.

Chouhan, TS (2002), Modern Techniques for Natural Disaster Management, ASTI Journal of Training and Development, Vol. III, No. 2, July-Dec., 43-78.

Chowdhury A and AB Majumdar (1983), Now westers and synoptic climatology of the hot weather season in north India, Vayumandal, Vol. 13, Nos. 1 and 2, Jan-Jun.

Chowdhury A and AK Banerjee (1983), A study of hail-storms over north-east India, Vayumandal, Vol. 13. Nos. 1 and 2, Jan-Jun.

Chowdhury, A and HR Ganeshan (1981), Geographical distribution of physioclimatic regims over India, Mausam, Vol. 32, No. 4, pp. 349-56.

Dasgupta SP (2002), Some tenets of Ancient Indian Agricultural Meteorology, Man and Environment, Centre for Study of Man and Environment, Calcutta Vol. 13.

Dasgupta SP (2002), The Indian concept of monsoon circulation, Geographical Review of India Vol. 64 (3).

Deosthali V (2000), Rationalizing agro-climatic zones of Maharashtra, Geographical Review of India Vol. 62(4).

Deosthali V and Iyer V (1996), Validation of sinusoidal model for simulation of diurnal temperatures, Transactions, Indian Institute of Geographers, Vol. 18.

Deosthali V and Payyappallia (1997), Climatological study of severe cold waves and prediction of minimum temperature over niphad, Geographical Review of India, Vol. 59 .

Deosthali V. (1997), Rainspell characteristics over Pune, The Deccan Geographer, Vol. 35.

Deosthali V. (1998), Sustainability of sorghum in Western Maharashtra scarcity zone, Indian Journal of Dryland Agriculture Research and Development, Vol. 13.

Deosthali V. (1999), Assessment of impact of urbanization on climate an appellation of bio-climateic index, Atmospheric Environment, Vol. 33.

Department of Agriculture and Cooperation, Ministry of Agriculture, Government of India (2001), Manual on Natural Disaster Management in India, National Centre for Disaster Management, Indian Institute of Public Administration, New Delhi.

Department of Disaster Management and Relief, Government of Rajasthan (2006), Annual Administrative and Progress Report 2005-06.

Department of Disaster Management and Relief, Government of Rajasthan (2004), Drought Management Manual.

Department of Disaster Management and Relief, Government of Rajasthan (2002), The Drought of 2002, Response and Management in Rajasthan.

Department of Science and Technology, Government of Rajasthan (1994), Resource Atlas of Rajasthan.

Dhar ON, AK Kulkarni and GC Ghose (1978), Rainfall distribution over Indian sub-division during the wettest and driest monsoons of the period 1901-1960, Hydrol. Sci. Bull. Vol. 23 No. 2 Jun.

Dhar ON, PR Rakhecha and AK Kulkarni (1982), Fluctuations in the north-east monsoon rainfall of Tamilnadu, Jour. Clim, Vol. 2, No. pp. 339-45.

Dhar ON, PR Rakhecha and BN Mandal (1978), Summer Monsoon rain in India and cyclonic distribution, Weather, Vol. 33, No. 5, May.

Dhar ON, PR Rakhecha and BN Mandal (1980), Is the number of cyclonic distribution traversing India during the monsoon season related to the rainfall in that season? Mausam, Vol. 31, No. 1.

Dhar ON, PR Rakhecha and BN Mandal (1980), Rainfall of Karnataka and its sub-division- a brief appraisal Ind. Jour. Power and Riv. Vall. Dev., Vol. 30, Nos. 5 and 6 May.

Dhar ON, PR Rakhecha and BN Mandal (1981), Influence of tropical disturbances on the monthly monsoon rainfall of India, Mon. Wea. Rev. 109, No. 1.

Directorate of Economics and Statistics, Rajasthan (2002), Some Facts About Rajasthan.

Directorate of Economics and Statistics, Rajasthan (2004), Basic Statistics- Rajasthan.

Directorate of Economics and Statistics, Rajasthan (2005), Statistical Abstract 2003.

Directorate of Economics and Statistics, Rajasthan (2006), Some Facts About Rajasthan.

Directorate of Economics and Statistics, Rajasthan, Jaipur (2002), Statistical Abstract, Rajasthan.

Directorate of Economics and Statistics, Rajasthan, Jaipur (2004), Basic Statistics, Rajasthan.

Drought Cell, Govt. of Karnataka (1995), Natural Disasters: Causes and Management Drought Monitoring Cell, Bangalore.

Drought in Rajasthan in National Seminar on Water Conservation and Drought Proofing of Arid Areas and its people, Social Policy Reserch Institute, Jaipur.

Drought Monitoring Cell, Govt. of Karnataka (2000), Drought Assessment and Response System: Chitradurga District, Karnataka, Drought Monitoring Cell, Bangalore.

Faroda AS, NL Joshi, S. Kathju and Amal Kar (ed) (1999), Management of Arid Ecosystem, Arid Zone Research Association of India, Scientic Publishers, Jodhpur.

Gilbert F. White (1974), Natural Disasters Local, National, Global, Oxford University Press, New York.

Gole S (1997), Geochemical mass balance in a small watershed in the seasonal propical region of the Western Ghats: The case of tributary in the source region of Koyna, India. Transactions, institute of Indian Geographers, Vol. 19.

Goswami DC (1980), Pluvial regime and flood hydrology of the Brahamputra river, In Kale VS (ed) Flood studies in India, Bangalore, Geological Society of India, memoir 41.

Goswami DC, Goswami ID, Duarah BP and Deka PP (1996), Geomorphological mapping of Assam using satellite remote sensing techniques, Indian Journal of Ceomorphology, Vol. 1 pp 225-35.

Government of Rajasthan (1995), A Brief Note on Development Perspectives in Rajasthan, Government of Rajasthan.

Government of Rajasthan (2003), Drought Situation and its Management in Rajasthan.

Gupta BL and Nityanand Singh (1981), Some findings of rainfall characteristics in Rajasthan, Geogr. Rev. Ind. Vol. 43, No. 4, Dec.

Gupta, LC and Vinod K.Sharma (2001), Drought in Rajasthan, 1999-2000 and 2000-2001, National Centre for Disaster Management, Indian Institute of Public Administration. New United Process, New Delhi.

Gupta, LC and VK Sharma (2001), Drought in Rajasthan 1999-2000 and 2000-2001 National Centre for Disaster Management, Indian Institute of Public Administration, New Delhi.

Gupta, Mukesh (2000), A new approach to long term solutions for drought proofing of arid area and its people in National Seminar on Water Conservation and Drought Proofing of Arid Areas and its people, Social Policy Reserch Institute, Jaipur.

Hema Malini B (1997), Global warming and climate change- Predictive models for temperate and tropical regions, the Deccan Geographer, Vol. 35.

Hemamalini B (1981), Patterns of aridity, spread and severity of droughts in Andhra Pradesh. The Ind. Geogr. Jour. Vol. 56, No. 1, Jun.

Indian Space Research Organization, Bangalore (1983), Disasters in India: Can Remote Sensing do something.

Ingle SR (2002), A comparative analysis of El-Nino and Rainfall over India and Goa- 1901-2000, Transactions, Institute of Indian Geographers, Vol. 24 (1 and 2).

Jagannathan P, HN Bhalme and PR Rakhecha (1978), Fluctuations of climate over Rajasthan, Prod. Ind. Nat. Sci. Acad. Vol 44 (B), No. 6.

Jain, Usha and CV Dhabriya (2000), Drought and Famine Elimination Scientific Strategy in National Seminar on Water Conservation and Drought Proofing of Arid Areas and its people, Social Policy Reserch Institute, Jaipur.

Jitendra Lal (1982), Is the Rajasthan desert spreading, Vayumandal, Vol. 12, No. 3 and 4, Jul. Dec.

Joshi VU and Kale VS (1997), Colluvial deposits in north-west Deccan India, Their Significance in the interpretation of late quarternary history, Journal of Quarternary Science, Vol. 12

Kalra, NK (2000), Planning drought combating and resources development of arid Rajasthan in National Seminar on Water Conservation and Drought Proofing of Arid Areas and its people, Social Policy Reserch Institute, Jaipur.

Kar A (1996), Morphology and evolution of sand dunes in the Thar Desert as key to sand control measures, Indian Journal of Geomorphology, Vol. 1.

Kar A (1998), Present day mobile crescentic dunes in the Thar desert India, In Alsharhan AS Glennie KW Whittle GL and Kendall, CGSC (ed) Quarternary Desert and Climatic change, Balkerna, Rotterdam.

Kar A (1999), A hitherto unknown palaeodrainage system from the radar imagery of south-eastern Thar Desert and its significance, Geological Society of India, memoir 42.

Kar A (1999), Climate and neotectonic controls on drainage evolution in the Thar Desert, In Paliwal BS (ed) Geological evolution of Nowth-west India, Scientific Publishers, Jodhpur.

Kar A (1999), Cluster analysis of grain size information for possible linkages between geomorphic environments: some case studies from Indian Arid Zone, Zeitschrift fur Geomorphologie, NF 43.

Kar A, Felix C, Rajguru SN and Singhvi AK (1998), Late Holocene growth and mobility of a transverse dune in the Thar Desert, Journal of Arid Environment, Vol. 28.

Kar A, Tsunekawar A and Miyazaki T (1998), Potential of global positioning system in sand due measurement: a case study from the Thar Desert, India, In Alsharhan, AS, Glennie, KW, Whittle GL and Kendall CGSC (ed) Quarternary deserts and climatic change, Balkerna, Rotterdam.

Kaul AK (1979), Seasons of Kashmir Valley Geogr. Rev. Ind. Vol. 41. No. 2, Jun. Pp. 123-30.

Kaur U (2000), Historical perspective of cyclonic storms and their impact in the Coastal Andhra Pradesh, annals of the National association of Geographers, India, Vol. XX(2).

Mahanti AC (1981), Origin of tropical disturbances, Archiv. Met. Geophys. Bioklim. Ser. A, 30.

Malik, Deepak (2000), Drought Proofing in National Seminar on Water Conservation and Drought Proofing of Arid Areas and its people, Social Policy Reserch Institute, Jaipur.

Mooley DA (1975), Climatology of the Asian summer monsoon rainfall (1) Control and concentration, Geogr. Rev. Ind. Vol. 37, No. 1, Mar.

Mooley DA (1976), Climatology of the Asian summer monsoon rainfall (2) patterns of monthly monsoon activity and their geographic distribution, Geogr. Rev. Ind. Vol. 38, No. 1, Mar.

Mooley DA (1976), Climatology of the Asian summer monsoon rainfall (3) some measures of central tendency, Geogr. Rev. Ind. Vol. 38, No. 3. Sept.

Mooley DA (1976), Climatology of the Asian summer monsoon rainfall (4) stability of the mean and median rainfall, Geogr. Rev. Ind, Vol. 38 No. 4, Dec. pp. 321-29.

Mooley DA (1976), Worst summer monsoon failures over the Asiatic monsoon area, Proc. Ind. Nat. Sci. Vol. 42, pt. !, No. 1, Jan.

Mooley DA (1977), Climatology of the Asian summer monsoon rainfall (5) space and time variations, Geogr. Rev. Ind. Vol. 39, No. 1 Mar.

Mooley DA and B Parthasarathy (1982), Fluctuations in the deficiency of summer monsoon rain over India and their effect on economy, Archiv. Met. Geophys. Bioklim. Ser. B, 30, 4.

Mooley DA and B Parthasarathy (1983), Variability of the Indian summer monsoon and tropical circulation features Mon. Wea. Rev. Vol. III No. 5, May.

Mooley DA and CM Mohile (1983), Study of cyclonic storms incident on the different sections at the coasts around Bay of Bengal, Mausam, Vol. 34, No. 2, April.

Mooley DA, B Parthasarathy and AA Munot (1981), Large-scale droughts over India and their impact on economy during the period 1871-1978 Annals Nat. Assn. Geogrs. Vol. 1 No. 2 Dec.

Mooley DA, B Parthasarathy, AR Subramaniam and GSLHV Prasad Rao (1981), Eco-climates of Rajasthan state, Geogr. Rev. Ind. Vol. 43 No. 1, Mar.

Mooley DA, B Parthasarathy, NA Sontakkke and AA Munot (1981), Annual rainwater over India, its variability and impact on economy. Jour. Clim. Vol. 1, No. 2, Apr.-Jun.

Mooley DA, VR Deshpande (1980), Rainfall probabilities in Karnataka during summer monsoon, Geogr. Rev. Ind. Vol. 39, No. 2, Jun.

Mukherjee AK and KP Padmanabham (1980), Rainfall over Bombay associated with cyclonic storms striking the coast, Mausam, Vol. 31 No. 1.

N. Jayanthi and N. Veeraraghavan (1984), Motion climatology and cyclonic storms over Indian seas, Vayumandal, Vol. 14, Nos. 3 and 4, Jul.-Dec.

Pant GB (1979), Role of tree ring analysis and related studies in palaeo-climatology: Preliminary survey and scope for the Indian Region, Mausam, Vol. 30, No. 4, Oct.

Pant GB (1983), Climatological signals from annual growth ring of selected tree species of India, Mausam Vol. 34, No. 3.

Parthasarathy B, and ON Dhar (1976), Study of trends and periodicities of rainfall over Madhya Pradesh. Proc. Ind. Nat. Sci. Acad. Vol. 42, pt. A, No. 1, Jan.

Parthasarathy B, and ON Dhar (1976), Study of trends and periodicities in the seasonal and annual rainfall of India, Ind. Jour. Met. Hydrol. Geophys. Vol. 27, No. 1.

Planning Department, Government of Rajasthan (2007), Draft Eleventh Five Year Plan 2007-2012.

Prasad SK and Pawar BC (1982), Discomfort over Bombay during pre-monsoon and post- monsoon months, Vayumandal, Vol. 12, Nos. 3 and 4, Jul.-Dec.

Pratap Narain and K D Sharma *et al.* (2000), Drought proofing in the Arid Zone of India in National Seminar on Water Conservation and Drought Proofing of Arid Areas and its people, Social Policy Reserch Institute, Jaipur.

Ram Mohan HS and VP Subrahmanyam (1980), Drought climatology of the semi-arid zone of Tamilnadu, the Ind. Geog. Jour. Vol. 55, No. 1, June.

Ram Mohan HS and VP Subrahmanyam (1983), Application of water balance concepts in agricultural planning and operation in India. The Nat. Geog. Jour. Ind. Vol. 29, pts 1 and 2, Mar-Jun.

Rama Rao K (1977), A new approach of computing discomfort index. Nat. Geogr. Jour. Ind. Vol. 23 pts. 3 and 4, Sept.-Dec.

Ramakrishna YS, RP Singh and KC Singh (1982), Influence of systems of planting pearl millet (Bajra) on crop micro-climate. Annals of Arid zone Vol. 21 No. 3, Sept.

Ratnam BP and SN Joshi (1983), Drought frequency for Rabi season at Dharwad and Bijapur. Annals of Arid Zone, Vol. 22, No. 1, Mar.

Ratnam BP, MM Hosmani and SN Joshi (1981), Water deficiency in Dharwad district. Annals of Arid Zone Vol. 20, No. 4 Dec.

Rawat JS and Rai SP (1997), Pattern and intensity of erosion in the environmentally stressed Khulgad watershed, Kumaun Himalaya, Journal Geological Society of India, Vol, 50.

Rawat JS, Rawat G and Rai SP (1996), Watershed hydrology under varied ecological conditions in Central Himalaya, National Geographer. Vol. 31.

RB Singh (1994), Space Technology for Disaster Monitoring and Mitigation in India, International Dentre for Disaster Mitigation Engineering, Institute of Industrial Science, University of Tokyo, Japan.

Reddy SS and Bh. V. Ramanamurty (1978), Thirty day periodicity in sun spot activity and rainfall. Ind. Jour. Rad, sp. Phys. Vol. 7, August.

Reddy SS, BK Mukherjee and Bh. V. Ramanamurty (1976), Solar activity and rainfall patterns in Tamilnadu during the north-east monsoon. Ind Jour. Rad. Sp. Phys. Vol. 5.

Roy NB and AK Mittal (1982), Rainfall distribution in the Union territory of Delhi. Vayumandal, Vol. 12, Nos. 3 and 4, Jul- Dec.

Sambasiva Rao M and K. Kalavathy (1983), Water balance and cropping patterns in Madurai district, The Ind. Geogr. Jour. Vol. 58 No. 2, December .

Sastri ASRAS, YS Ramakrishna and CS Bhandari (1982), Drought climatology of Gerhiya catchment in Rajasthan. The Nat. Geogr. Jour. Ind. Vol. 28 pts 3 and 4 Sept.-Dec.

Sharma, S N (1993), Drought in Rajasthan in Natural and Human Resources (ed.) T S Chouhan, Scientific Publisher, Jodhpur.

Shrimali, ML (2000), Empowerment of the people to Combat Drought and water conservation in Arid/Semi-arid areas in National Seminar on Water Conservation and Drought Proofing of Arid Areas and its people, Social Policy Reserch Institute, Jaipur.

Sikka DR (1982), Climatological study of tropical storms of the south Indian Ocean. Jour. Mar. Biol. Assn. Ind. Vol. 17 No. 3 .

Singh S (1997), Drifting sand hazard and its control in the Thar Desert, Indian Journal of Geomorphology, Vol. 2.

Singh S and Dubey A (1996), Geographic significance of ambrothermic conditions of anthropogenically manipulated gully basin of sub-humid tropical environment, National Geographer, Vol. 31.

Singh T (2000), Climate variability and its impact on crop productivity: A case study of Bihar Plateau, India, Proceedings of International symposium on Climate Variability and Their Impact, Commmission on climatology, 29th IGC, Seoul, Korea.

Singh T (2000), Drought- A challenging natural hazard in sub-humid climatic region: A case study of Palamau in RB Singh (ed) Disaster management, Rawat Publications, New Delhi.

Singh T (2000), Drought and El-Nino, Down to earth, Vol. 8(24).

Singh T (2001), Climate variability and change: Indian scenario, annals of the National Association of Geographers, India Vol. XXI (2).

Social Policy Research Institute, Jaipur (2003), A Compendium of Papers presented in National Seminar on Water Conservation and Drought Proffing of Arid Areas and its people.

Subrahmanyam VP (1967), The problem of incidence and spread of continental drought, Technical note No. 2, WMO/ IHD Series, Geneva.

Subrahmanyam VP and P Viswanadham (1979), A study of rainfall patterns in Visakhapatnam district, Ind. Geogr. Jour. Vol. 54 No. 2, Dec.

Subrahmanyam VP, AALN Sarma (1980), Evolution of Indian climates and water use planning Trans IIG Vol. 2, No. 1, Jan.

Subrahmanyam VP, AALN Sarma (1981), A climatic study of human comfort zones in south India. Proc. Internat. Symp. On human health and tropical diseases (IGU) Dept. of Geogr. University of Madras, December.

Subrahmanyam VP, HS Ram Mohan (1980), Classification of monsoon climates and stability of their moisture regime in Statistical Climatology, Development of atmospheric science 9130 (ed) S Ikeda *et al.* Elsevier Publ. Co.

Subrahmanyam VP, Sachi Devi and P. Viswanadham (1982), Some aspects of urban climatology of Visakhapatnam. Trans. IIG, Vol 4, No. 2, July.

Subrahmanyam VP, VM Sastry and K Sachi Devi (1978), A comparative study of temperature and vapour pressure distributions in urban and non-urban locations, Geogr. Rev. Ind. Vol. 40 No. 1, March.

Subramaniam AR and A Sambasiva Rao (1982), Water balance and its applications to Maharashtra State. The Nat. Geogr. Jour. Ind. Vol. 28. pts. 3 and 4, Sept.-Dec.

Subramaniam AR and A Sambasiva Rao (1984), Agroclimatic classification of Maharashtra. Trans. IIG, Vol. 6 No. 1, Jan.

Subramaniam AR and GSLHV Prasada Rao (1981), Ecoclimates of Rajasthan State, Geogr. Rev. Ind. Vol. 43. No.1, March

Subramaniam AR and M Benkateswara Rao (1983), A climatic study of drought types in Rayalseema, Geogr. Rev. Ind. Vol. 45, No. 2, Jun.

Subramaniam AR and PVSSK Vinayak (1982), Water balance studies of Trivendrum and its vicinity, Trans IIG, Vol. 1, Jan.

Subramanyam VP (1982), Drought climatology, Annals of NAGI Vol. 1, Jan.

Subramanyam VP (1982), Esstern Ghats vegetation and climatic aspects, Proc. Nat. Sem. On Resources, development and environment of the Eastern Ghats, Andhra University, Waltair, March.

Subramanyam VP (1983), Applied climatology, Vol. 3 of contributions of Indian Geography, Heritage Publishers, New Delhi.

Subramanyam VP (1983), General Crimatology- Vol. 3 Contributions to Indian Geography, Heritage Publishers, New Delhi.

Subramanyam VP (1983), Some aspects of hydro-climatology and water balance of the Chandrampalem basin, Trans IIG Vol. 5 No. 1 January.

Subramanyam VP, HSG Ram Mohan and Pardhasaradhi (1979), Appraisal of ground water potential of the dry climates of India- a climatic approach. The Nat. Geo. Jour. Ind. Vol. 25 pt. 1 March.

Swadesh Mishra (1983), Agricultural drought over Gengatic West Bengal in relation to Anaan rice, Geogr. Rev. Ind. Vol. 45 No. 2 June.

Syiempieh H and Das P (2003), Rainfall trends in the Brahmaputra Valley of Assam, in Z. Hussain (ed) Environmental Issues of North East India, Regency Publications, New Delhi.

Tali BM and HM Qureshi (1979), Rainfall trends over southern Rajasthan, Geogr. Rev. Ind. Vol. 41 No. 4 December.

Thornthwaite CW (1947), Climate and moisture conservation, Annals, Association of American Geographers, Vol. 37. No. 2.

Thornthwaite CW (1948), Qn qpproqch towards a rational classification of climate. Geographers Review, Vol. 38. No. 1.

Thornthwaite CW and Mather JR (1955), The water balance. Publns. In Clim. Drexel Instt. Tech. Vol. 8. No. 1.

Tiagi SSL (1980), Climatology of the foot region of Shivalik hills, Mausam, Vol. 31, No. 2.

Tiwari AK (ed.) (1988), Desertification: Monitoring and Control, Scientific Publishers, Jodhpur.

Tsunekawar A, Kar A, Yanai J, Tanaka U and Miyazaki T (1997), Influence of continuous cultivation on the soil properties affecting crop productivity in the Thar Desert of India, Journal of Arid Environment, Vol. 36.

Tsunekawar A, Miyazaki T and Kar A (1996), Desert mapping using NOAA/ AVHRR in Rajasthan, India, In singh RB (ed) Disaster, environment and development, Oxford and IBH Publishing co. Pvt. Ltd. New Delhi.

UNESCO (1977), Development of Arid and Semi-arid Lands: Obstacles and Prospects.

Venkateshwarlu J. and Kar A. (1996). Wind erosion and its control in arid North-west India, Annals of Arid Zone, Vol. 35.

Verma, BL (1978), Droughts in Rajasthan, Department of Geography, University of Rajasthan, Jaipur (Ph.D. thesis unpublished).